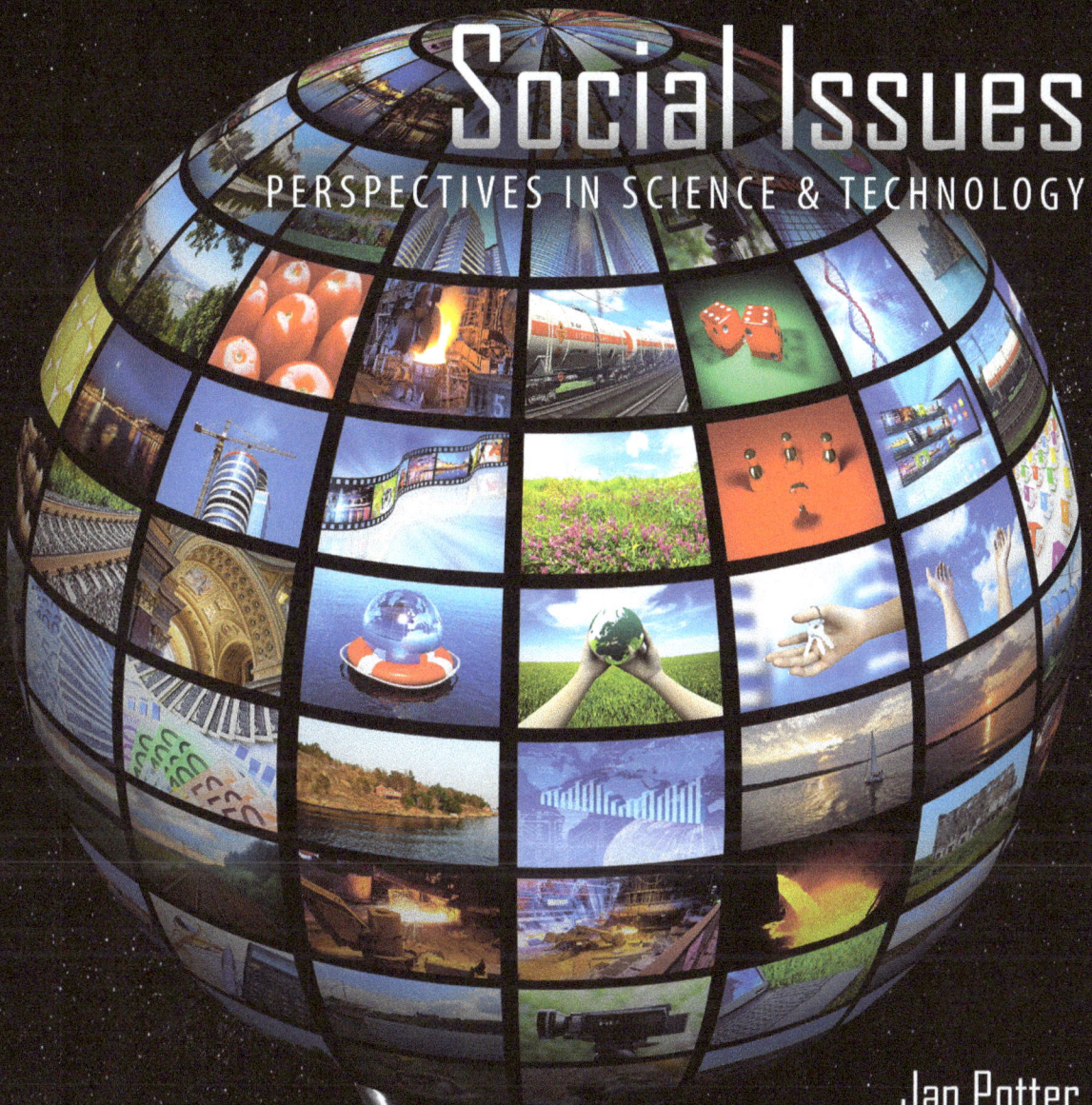

Social Issues

PERSPECTIVES IN SCIENCE & TECHNOLOGY

Jan Potter

Trina Queen

Kendall Hunt
publishing company

Cover image © Shutterstock, Inc.

Kendall Hunt
publishing company

www.kendallhunt.com
Send all inquiries to:
4050 Westmark Drive
Dubuque, IA 52004-1840

Copyright © 2015 by Kendall Hunt Publishing Company

ISBN 978-1-4652-9320-6

Printed in the United States of America

Contents

Preface

"Your brain may give birth to any technology, but other brains will decide whether the technology thrives. The number of possible technologies is infinite, and only a few pass this test of affinity with human nature."

—Robert Wright, Nonzero: The Logic of Human Destiny[1]

Why take a class about the interaction of science, technology, and society? Many of the educational institutions today are concerned with teaching you the facts you need to succeed in your future career. If you want to study chemistry, then you will learn what you need to know about that field. If you want to study mathematics, then your classes will be filled with equations, problem solving, and concepts.

If you want to study great literature, or even how to make a video game, you will most likely learn the basics for each of those fields. The problem is that you will not, unfortunately, live in a world that is full of people who only do those things. The real world turns on a synthesis of everything. You might have the best idea in the world for a product that is amazingly time-saving and beautifully designed. However, you might know nothing about marketing or sales. You might not have learned that someone else had the same idea. Ideas, while wonderfully innovative, rarely ever produce real "things."

Most people realize that technology can shape society. The rise of cell phones has changed how we interact with each other in ways that few people predicted 10 years ago. Facebook has become a social construct that was never even imagined in the past. It is important to realize, though, that society shapes technology as well. Human behavior, gender, media, politics, and religion can all play a significant part in whether or not a new idea becomes the norm.

In this textbook, we will look at how things interact and succeed or fail. Through the lens of society and culture, we can perhaps see what the future will bring.

---

[1] Robert Wright, *Nonzero: The Logic of Human Destiny* (New York: Vintage Books, 2001), 27.

Science Fiction: The Creative Roots of Technology

Science fiction is a particular literary genre that has often been used to predict future technological marvels. Sometimes it is very difficult to ascertain whether a science fiction story provides the incentive to design a new and fantastical "object" or if the story is created based upon the "what-ifs" inherent in the actual design. It is true, however, that NASA, for example, has hired and consulted with science fiction writers in the past. Science fiction can provide a creative solution or simply a new idea to study.

Good science fiction almost always has some degree of scientific or technical "believability" in the stories. The beauty of the genre is that even the minutest detail might give another scientist an idea to explore.

We definitely owe a debt to early science fiction authors like Jules Verne and H. G. Wells. Both are often considered the founding fathers of this sort of literature. The industrial revolution not only brought about new technologies, it also produced a new way of looking at the world. Science and technology became a new kind of god that we imagined would solve all of our social ills.

Jules Verne was one of the most popular authors in many years. He was born in Nantes, France, in 1828 and died in Amiens, France, in 1905. Throughout his life, he was an extremely popular author. He and H. G. Wells are generally considered the founding fathers of science fiction literature.

Verne was said to be influenced by Victor Hugo and Alexandre Dumas; this gave him the idea that science could be both intriguing and heroic. He wrote his first science fiction story in 1854 after reading a French translation of Edgar Allen Poe. The story called "The Sphinx of the Ice Fields" was written in response to a strange Poe story called "Narrative of Gordon Pym." Neither book was a commercial success.

Verne was not a scientist or a world traveler—but you would never know that reading his books. He is known for an amazing ability to research information. He was particularly hostile to authors like Lewis Carroll; Verne thought that the concept of fantasy literature was offensive to intelligent people. He believed that good stories should be very detailed and correct as to even the smallest of details. He was appalled when H. G. Wells "invented" an antigravity substance called "cavourite."

I sent my characters to the moon with gunpowder, a thing one may see every day. Where does M. Wells find his cavourite? Let him show it to me![1]

Verne believed that any stories written should be correct with the current scientific discoveries of the day. However, in many of his stories, he also bent his stories to follow a plot line that would probably not be considered acceptable science. His work showed that science and scientific discovery could be fascinating reading when coupled with adventure.

His novel "Journey to the Center of the Earth" is not heavy on science, although it could be argued that we did not know that much about Earth's core when the novel was written. Studies had been done by the mid-nineteenth century that questioned the origin of Earth as being of religious origin and that laid the groundwork for future geological studies. Verne's story begins with a "gentleman scientist" who discovers a book of ancient Icelandic runes, which he then translates. It basically tells how to reach the center of Earth from a point in Iceland. The idea behind it was that explorers would use a volcanic crater to travel down to Earth's core. In the novel, this was done with little danger. At the end of their travel through Earth's core, the explorers found a land full of prehistoric life and bizarre phenomena. Once Hollywood decided to film the story, the land became the Lost City of Atlantis.

Verne wrote *Twenty Thousand Leagues Under the Sea* in 1869. Arguably his most scientific work, it included many of the major themes you could find in Literature 101: man versus nature, man versus man, and man versus himself. Unfortunately, the first English translation in 1873 left off about a fourth of the original work and it included many errors of translation (including the fact that Verne's title was: "Twenty Thousand Leagues Under the Seas." It was nearly a 100 years later before Verne's work was more correctly translated into English.

The story of the book is that people were seeing some kind of "sea monster" and a formal expedition was sent out to discover just what it was. What they discovered was a submarine, which was captained by the illusive Captain Nemo. Originally, Verne wanted Captain Nemo to have a more specific political agenda (the original idea was that he was Polish), but, after much argument, he agreed to leave Nemo's background to a more "unknown" allure.[2]

Captain Nemo comes across as a kind of antihero. Verne clearly saw that technical advancement would be the future of the new "scientific" world. Nemo ruled at the top of a food chain as a scientist, inventor, engineer, and genius. He was autocratic, dark, and somewhat obsessed. This was part of the rise of the idea that science would save us all—the world would (and should) be run by engineers. In the story, he is shown next to characters like Professor Aronnax, who represents the "old" way of book learning in a school environment.

The submarine itself was a design marvel. Clearly delineated in the book, Verne included very clear information about how the ship was made and what powered it (electricity, which was a new modern marvel of the times). The Nautilus (a name taken from Robert Fulton's first "successful" submarine built in 1800) was a huge mechanical construct that terrified people by its "sea monster" shape.

There are interesting correlations between the design of the Nautilus and the Civil War submarine, the H. L. Hunley. The Hunley was used in several skirmishes during the Civil War in the United States. In 1864, it became the first submarine to actually sink another vessel. Unfortunately, it also sank and all on board were killed. The wreck was raised in 2000 from Charleston Harbor. Since it was built before the publication of *Twenty Thousand Leagues Under the Sea,* there is some possibility that Verne had done research on the concept before the Hunley was built.

---

[1] "Jules Verne: A Man Before his Time," *ROH Press*, accessed December 19, 2014, http://www.rohpress.com/verne.htm.

[2] Arthur B. Evans, "Hetzel and Verne: Collaboration and Conflict," *Science-Fiction Studies #83* 28, no. 1 (2001): 97–106, accessed December 9, 2014, http://www.depauw.edu/sfs/review_essays/evans83.htm.

The book also included such scientific marvels like underwater breathing apparatus. While a version of this had been invented by others, Verne takes it one step further to make an untethered version. Once the book became fodder for the movies of the twentieth century, much of the original technical detail was lost. It does give us a good starting point for the use of the science fiction genre as it relates to a study of the interaction of science, technology, and society. It also gives us the beginning of how the interaction of the three can affect how we view science in our daily lives.

The other modern "founder" of science fiction is Herbert George Wells (H. G. Wells).

H. G. Wells was an English novelist, journalist, sociologist, and historian. He was born in 1866 and died in 1946. His father was a cricket player, and, in 1877, he broke his leg and lost the majority of his income. H. G. was brought home from school and made to enter an apprenticeship as a draper. He was not overly successful in that field and for some time followed other apprenticeships equally as unsuccessful. He was fortunate, however, in that he had access to good libraries which interested him in writing and literature. He spent time as a teacher and finally acquired teaching credentials. He married his cousin, Isabel, whom he later set aside when he fell in love with one of his students.

© Courtesy of Library of Congress

He is generally credited with writing the first "modern science fiction" stories and, at the time of his death, had written over 100 books (half of which were fictional). Wells referred to his work as "scientific romances." Many items in his work went on to become real inventions in later years.

War of the Worlds was written in 1898 and was originally put forth as a response to the militarization of Germany at that time. In 1894, Mars had been very close to Earth and the idea of little Martians coming to visit was a newsworthy item. An Italian astronomer named Giovanni Schiaparelli reported seeing "canali" on Mars (allegedly he meant channels), but the popular press reported it as canals. People claimed to see lights on Mars. It was a popular novel of the time.

Wells' *War of the Worlds* became even more famous when it was performed as a radio play by Orson Welles Mercury Theatre on October the 30th of 1938. The radio production had a very limited audience, but the "staging" of the radio play was very unique. At that time, there were no mandatory requirements that a radio station had to identify itself at certain intervals (in fact, that requirement came partly as a result of the furor over this radio play). At the beginning, the (small) audience was told that it was only a play—a dramatization of the book. However, there were no commercials or any other kind of breaks during the performance as there would be today. It did not follow the book; instead, it was done as a "breaking news" format. It began with ballroom dance music from a famous New York dance floor and then the announcer "broke in" to tell the audience that something had happened. It was done in very high drama (with repeated "break-ins" to the dance music) until a fever pitch of excitement existed. There were (fake) interviews with famous astronomers and faculty. It seemed very real.

Normally, such a live theater play would have a very small audience. The most popular show of the time was the Charlie McCarthy Show (why a ventriloquist show would be the most popular radio show is a curiosity in itself). It being a more "modern" and well-financed show, it did have commercial breaks. Audiences then (as they do now) began idly flipping radio stations during the break. Unfortunately, they stumbled into the middle of the *War of the Worlds* radio play. Many people believed it to be an actual recounting of an actual Martian invasion.

One of the four guardian gargoyles over Pont del Regne (bridge of the kingdom). Valencia
© Shutterstock/Alfonso de Tomas

H. G. Wells is also famous for *The Time Machine*. Published in 1895, the book details the story of a "dilettante" scientist who builds a time machine. He uses the machine to go back and forth in time. The books were actually written to talk about the English class divisions of the time, but it became one of his more famous science fiction works. The narrator, Professor Hillyer, goes back in time and returns to tell his friends all about it. He ends up accidentally traveling to the year 802701 where he finds the curious "future" of humankind. There are two types of people—the Eloi who live above ground and spend their time all day laughing and playing. They do not work, and they are uncommonly beautiful. They are the antithesis of the "Morlocks," who basically represent the industrial cannon fodder of the time—the underlying class of people who do all of the work.

However, as a cautionary science fiction tale, these Morlocks "raise" the Eloi as a food source. They are portrayed as ugly villains. Wells was somewhat of a fan of Darwin, and he seemed to be implying that the future might not end well if the English leisure class continued to play and do no viable work. In the novel, the Eloi are shown as having no learning; they cannot read or write. What first seems appealingly beautiful (a life empty of toil and strive) is actually an indication that that things must change. Wells was a sociologist; he believed that some pain and suffering are necessary to human growth.

The most popular movie version of this story followed most of the precepts of the story. A later version was filmed in 2002, directed by Simon Wells, a grandson of H. G. Wells. The premise of the story was changed; S. Wells allegedly did not feel that a struggle based upon labor models would be successful in today's world. Instead, he created a version where the "death" of society as we know it was brought about by technology.

Another of H. G. Wells famous works was *The Island of Dr. Moreau* published in 1896. This book tells the story of a medical doctor who decides to cross-breed various animal species (including humans) to produce desired characteristics. You can see how this progressed when we look at the concept of chimera research in later chapters.

H. G. Wells is also credited with many predictions or scientific "hits" of technology that would come to shape the future. In his 1902 book, *Anticipations of the Reaction of Mechanical and Scientific Progress Upon Human Life and Thought,* he discusses the effect of new railway systems on how we will move about in the future. He pretty clearly predicts the effect of modern subway systems.

When he discusses such technological achievements, he surmises that the availability of rapid and reliable transportation would eventually mean that people would be moving away from cities toward more "natural" sites. He speaks extensively of the future roles of people (especially of the idle rich and the poor laborers) and predicts that moral restrictions would decline as we have more sexual freedom:

> "It is foolish, in view of all these things, not to anticipate and prepare for a state of things when not only will moral standards be shifting and uncertain, admitting of physiologically sound ménages of very variable status, but also when vice and depravity, in every form that is not absolutely penal, will be practiced in every grade of magnificence and condoned."[3]

---

[3] Herbert G. Wells, *Anticipations of the Reaction of Mechanical and Scientific Progress Upon Human Life and Thought* (London: Chapman and Hall, 1902), 134, http://www.gutenberg.org/files/19229/19229-h/19229-h.htm.

In his novel, *The Sleeper Awakes*, he talks about doors that open automatically. In *The Shape of Things to Come*, he talks about wireless communication devices.

He speaks a good bit about war and how it has changed with newer and better technologies. Where once we had cavalry and sabers, rifles would become more precise and change how war was conducted. He speaks of how such rifles might be connected to balloons (Anticipations, pp. 179–183). He discussed the ramifications of tank warfare (*The Land Ironclads*, 1903) and "aerial" campaigns (*The War in the Air,* 1908) and how technology will change how we wage war.

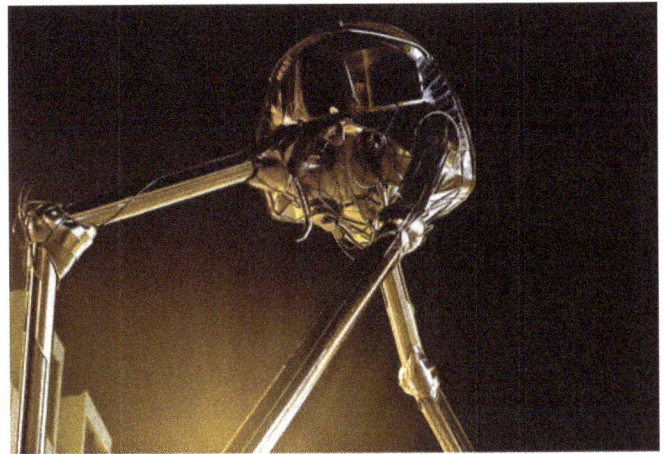

This sculpture celebrates Woking as the birthplace of H.G. Wells' novel The War of the Worlds.
© Shutterstock/Zryzner

His "heat ray" in *The War of the Worlds* functioned much as a laser weapon would today. It is a "handheld" device that can destroy whatever it touches. In this novel, he also included a good portrayal of what would become industrial robots. While we don't have "real" ray guns used in warfare, it is not for a lack of trying. *Forbes* had an article in 2013 about a kickstarter program (which subsequently failed) for a science fiction story that was to fund a physics project to build a "ray gun."[4]

Some current historians have disputed just how great the impact was; but first-person reports tell a story of nearly mass hysteria. In fact, Orson Welles was briefly arrested but they realized that there was nothing he could be charged with; it was not a crime. This was one of the best examples of the early power of the media to incite behavior. It would not be the last.

One of his most amazing predictions first appeared in an early (and mostly forgotten) novel, *Tono-Bungay* (1909). It was later expanded in Wells' 1914 novel, *The World Set Free*. In these books, he outlines the process of radioactive decay and the possibility of using it as a bomb. In fact, Wells is credited with the first use of the phrase "atomic bomb" in his aforementioned novels. Leo Szilard acknowledged that Wells' book inspired him to discover the nuclear chain reaction.

[Rhodes, Richard, The Making of the Atomic Bomb, Simon & Schuster, 2012]

In his story, "The Stolen Bacillus," written in 1895, H. G. Wells tells us a story about a man who infects himself with a disease so that he can eradicate a population. He refers to it again in *War of the Worlds*. This idea has been around for many centuries (you can read about catapulting plague-infected bodies), but this story was chillingly prescient about what could happen in the future with biological warfare.[5]

---

[4] Michael Venables, "Ray Guns Beam from H.G. Wells to Kickstarter and Beyond," *Forbes* (blog), January 23, 2013 (10:11 p.m.), http://www.forbes.com/sites/michaelvenables/2013/01/23/ray-guns-kickstarter-and-beyond/.

[5] Herbert G. Wells, *Anticipations of the Reaction of Mechanical and Scientific Progress Upon Human Life and Thought* (London: Chapman and Hall, 1902), http://www.gutenberg.org/files/19229/19229-h/19229-h.htm.

His 1902 words on gender equality were prescient:

> "Now, it is difficult to say why we should expect the growing girl, in whom an unlimited ambition and egotism is as natural and proper a thing as beauty and high spirits, to deny herself some dalliance with the more opulent dreams that form the golden lining to these precarious prospects? How can we expect her to prepare herself solely, putting all wandering thoughts aside, for the servantless cookery, domestic Kindergarten work, the care of hardy perennials, and low-pitched conversation of the engineer's home?"[6]

and

> "In these Anticipations it is impossible to ignore the forces making for a considerable relaxation of the institution of permanent monogamous marriage in the coming years, and of a much greater variety of establishments than is suggested by these possibilities within the pale."[7]

He did get some things wrong, though. He was not a fan of Jules Verne.

"I must confess that my imagination, in spite even of spurring, refuses to see any sort of submarine doing anything but suffocate its crew and founder at sea."[8] (Anticipations, p. 201). And, while we do not have invisibility or time travel quite yet, that does not mean that it will never happen. Scientists have been working on the invisibility cloak for some time[9,10] and on HyperStealth Biotechnology[11].

© Shutterstock/oneinchpunch

One thing we do know is that scientific discovery and development build upon the stories we have heard. Whether they actually happen, however, depends upon the cultural climate of the times. Are we ready for time travel or invisibility? What would this mean for us as a society? Of course, the best argument that we will never manage time travel is, quite simply, that no one has ever come back to tell us about it. We are approaching at least the possibility of invisibility, however. There will be more about this later when we talk about war and scientific endeavors.

---

6 Ibid., 121.

7 Ibid., 126.

8 Ibid., 201.

9 Max Eddy, "5 Serious Attempts to Build an Invisibility Cloak," *PC Magazine*, December 15, 2012, http://www.pcmag.com/slideshow/story/306035/5-serious-attempts-to-build-an-invisibility-cloak.

10 "Quantum Stealth Invisibility Cloak, From Canada's Hyperstealth Biotechnology, Gets Pentagon Backing," *Huffington Post Canada*, December 19, 2012, http://www.huffingtonpost.ca/2012/12/11/quantum-stealth-invisibility-hyperstealth_n_2277394.html.

11 HyperStealth Biotechnology Corp., "Leaders in Camouflage, Concealment, and Deception," http://www.hyperstealth.com.

DISCUSSION QUESTIONS

1. What do you think about Verne's idea of scientists as the top of the food chain? What do you think about the idea that heroic figures could be without religion or what we customarily call a "moral" belief structure?

2. Using technology is important in Jules Verne's work. Do you think Verne sees both positive and negative aspects of science and technology? Do you think that most people see technology for good or evil?

3. Can you correspond the Eloi and the Morlocks with good and evil? What do you think about the idea that the rich could die out because they are physically weak (that physical weakness could lead to poor moral behavior)? One theory about the book is that pain and suffering are necessary for your mental and moral health. Agree or disagree?

4. Do you believe that time travel is possible? Do you think that some change you accidentally made in the past could have huge repercussions when (or if) you returned?

5. Do you think modern media could create the same sort of hysteria as there was in *War of the Worlds*? How?

LITERARY SOURCES—FICTION

(In addition to all of the books listed in the text)

Islands in the Sky: The Space Station Theme in Science Fiction Literature (I.O. Evans Studies in the Philosophy and Criticism of Litera) Paperback—July 10, 2009 by Gary Westfahl

RESOURCES AND REFERENCES

The Hunley:

http://www.onr.navy.mil/focus/blowballast/sub/history3.htm

http://hunley.org/index.asp

http://news.nationalgeographic.com/news/2012/01/pictures/120131-hunley-civil-war-first-submarine-science-nation/

http://www.navy.mil/navydata/cno/n87/usw/issue_21/verne.htm

http://io9.com/which-real-life-victorian-sub-inspired-jules-vernes-na-458345856

http://articles.baltimoresun.com/2000-02-11/news/0002110290_1_hunley-jules-verne-horace

Social Issues: Introduction

We have moved from a society that communicated via oral traditions through the evolution of written materials to an Internet-based society. Today, there is an argument that children no longer need to learn "facts" because they will most likely have instant access to any facts they need. In some ways we have come full circle. Critical thinking has become far more important than it once was.

Where we once built friendships upon a commonality of work or play, we have moved into the time when being a "friend" might mean nothing more than clicking on the Like button on an online post. Social media brings the mundane closer; we are learning about people in ways that may not prove useful (do you really need to know what someone had for dinner – with a picture?). It can be, however, an impressive way of sharing news – both public and personal to many different people at the same time. You can marshal a near instant response to any kind of information by a simple posting on a social media site. You can write a critical review, or post a simple message that you have waited X hours in an emergency room for help. The posting itself can both educate your family and friends and, in some cases, cause the hospital to respond to you more quickly.

Buying and shopping habits are changing; if you search for something online then you will see that item popping up all over your social media posts. Current events can be near instantly published. Our social construct is changing daily.

Perhaps this is good news. Perhaps it is not. Buy whatever happens, it is here to stay. The Internet has redefined "social" and we will have to adjust our expectations as well.

Social Issues: Human Behavior

"The real problem is not whether machines think but whether men do."

—B.F. Skinner

HUMAN BEHAVIOR

In the mid-eighteenth century, a philosopher named Jean Jacques Rousseau introduced the idea that man in a "state of nature" was inherently good. In other words, if there was no influence from society, culture, government, or anything else, we can define man as "good." It doesn't sound like a very novel concept, but many people, particularly Christians, believed at that time that man was born flawed from the concept of original sin. Man was considered weak and only by grace/religion could he learn to withstand temptation.

This introduced the idea of what would happen if a person were raised by animals and not by people. There are several incidences of this throughout history; most are simple legends like Romulus and Remus from Roman history.

Another is the "Wild boy" from Aveyron. This child, approximately 10 or so, wandered out of the woods and the story grew up around him that animals had raised him. Francois Truffaut made a famous movie about him called "L'Enfant Sauvage."

However, there have been a few somewhat documented studies of "found" children. One such example happened in Northern India. In 1920, a missionary observed two young girls climbing out of the den of wolves. The missionary decided they needed saving and took them home with him. He named them Kamala and Amala. The main problem with the story is that, by some accounts, there were no witnesses to the missionary's tale of ferociously fighting the wolf mother for the children. In addition, there is some evidence that he made money off the story as well. One of the children died very quickly, but the other lived several years longer.[1]

Romulus and Remus by Wenceslaus Hollar (1607–1677)
University of Toronto Wenceslaus Hollar Digital Collection

---

[1] Jun-Hao Rosalyn Shih, "Daughters of the Wild: The Feral and Human Perspective," The Morningside Review, 2007/2008, http://morningsidereview.org/essay/daughters-of-the-wild-the-feral-and-human-perspective/#.

Der Wilde von Aveyron.

Victor of Aveyron, 1801

The idea of a true "savage" raised by wolves is far more interesting than the reality. It is said that both girls never learned to walk upright or communicate effectively or be housebroken. It is more likely that these are children that were abandoned due to some specific handicap. However, the idea quickly grabbed the attention of the public.

ENTER TARZAN, NOBLE SAVAGE

Edgar Rice Burroughs was not the first person to capitalize upon this idea, but he was certainly one of the most prolific. He first published the story of Tarzan in 1912. It is the story of a baby whose parents died "in darkest Africa" and the baby was subsequently raised by great apes. There are approximately 24 books about Tarzan and his family. As it turned out, Tarzan's parents were actually English nobility and he was later found.

Burroughs was very interested in the ideas concerning heredity and environment. By the time of his death, his total sales from books were estimated between 30 and 60 million copies. He is considered to be "the most widely read American author of the first half of the twentieth century."[2]

To the city slicker people of the twentieth century, Tarzan was an epic and heroic figure. He feared nothing. His story was originally published as a "serial" novel. You could buy one "chapter" for a small amount and the story would end with some terrible "the tiger leapt" nonending. You would have to buy the next installment to see what happened to our fearless Tarzan. Speaking of tigers, the story takes place in Africa. A less than kindly reader wrote Burroughs to tell him that there were no tigers in Africa, so Burroughs changed the story to a lion.

While not the first, the Tarzan series was extremely popular and Burroughs held on to the copyright for many years. He also encouraged the merchandising of toys to go with the Tarzan story.

So, why did people living all over the world see Tarzan's behavior as heroic? What exactly is a hero? The cartoon Mighty Mouse always used the same tagline: "Here I come to save the day." F. Scott Fitzgerald said, "Show me a hero and I'll write you a tragedy." In the movie Serenity, Zoe Warren says "Do you know what the definition of a hero is? Someone who gets other people killed."[3]

For the ancient Greeks, all it took was to be a free man who fought in the Trojan War. The definition has changed over the years. In his book, "Hero's Journey," Joseph Campbell says that there is a certain cycle to the concept of heroes in literature (or life). He calls it the "monomyth."

1. A person begins as an innocent child.
2. He/she is called to some sort of adventure or "otherness" outside the routine life.
3. There is some sort of initiation.

---

[2] John Taliaferro, Tarzan Forever: The Life of Edgar Rice Borroughs, Creator of Tarzan, (New York, Scribner, 1999), https://www.nytimes.com/books/first/t/taliaferro-tarzan.html.

[3] Zoe Warren, Serenity, directed by Joss Whedon (2005; Universal City, CA: Universal Pictures), DVD.

4. There are various trials and a low point of despair.
5. There is atonement or resolution of belief.
6. There is, after this cycle, the freedom to be whole and real.

George Lukas used this version of "life" to design the Star Wars epics because, "There was no modern mythology to give kids a sense of values, to give them a strong mythological fantasy life. Westerns were the last of that genre for Americans. Nothing was being done for young people with real psychological underpinnings."[4] Does a modern society still need its heroes? What happens when we combine the heroic with the modern technological age?

Are All Movie Heroes the Same Person? Joseph Campbell's theory of the Monomyth suggests that all mythical and legendary tales told throughout human history share a common structure.[5]

Today, however, nothing is quite so simple. Today, if we hear of a heroic deed, we are immediately inundated with knowledge, rumor, backstory, and every kind of information. We will come back to this topic.

We will speak in a later chapter about the idea of monsters and how we treat and have treated them in history when we discuss the story of Frankenstein. The concept of how a modern society views and will view both heroes and monsters is connected. One thing that monsters can be said to have taught us is the concept of how we would react to such a thing. Think about how we react to a horror movie, for example. We can be terrified, but it does help us to learn how we would behave in a like situation. Stay away from open windows, run when you can, never turn your back, and so on. The idea here is that such horrific concepts help you to deal with things that scare you from a safer secondhand platform. You have probably already figured out that when the zombie apocalypse comes, you need to know all about the proper way to shoot them and/or to run faster than someone else. Hopefully, that is knowledge you will never need.

When we watch a horror movie or see an ordinary person perform something heroic, we begin to imagine how we would act in such a situation. Perhaps we need monsters in our lives. Many years ago, there were famous storytellers like Hans Christian Andersen who would relate chilling tales about what happened to children who did not behave well. The story of the Little Match Girl described a child freezing to death. The original Grimm's Fairy Tales were equally as "grim." The German stories about Struwwelpeter were astonishingly gory and difficult to see today. Bad children were viciously punished by a character known as "Struwwelpeter."

Heinrich Hoffmann's *Struwwlpeter* (1845)

[4] George Lucas, interview about George Lucas, American Masters, PBS, January 13, 2004, http://www.pbs.org/wnet/americanmasters/episodes/george-lucas/about-george-lucas/649/.

[5] Paul Hiebert, "Are All Movie Heroes the Same Person?" Pacific Standard, last modified June 20, 2014, http://www.psmag.com/books-and-culture/movie-heroes-person-joseph-campbell-monomyth-83796.

These types of materials are no longer available to children. We have sanitized our children's literature to such a degree that most unpleasantness has been removed. This tends to be cyclical, but such concepts are considered to be traumatic for infant brains.

Do we need this? From a Freudian point of view, it can be said that we need monsters because we all have a "bad" side—so killing a monster is a way of making us good again. It is a type of cathartic journey into our own subconscious.[6]

In like manner, our concept of heroes has been changing over time. After World War II, there was a general consensus that all was well and that, since we won, we were perpetually going to be the wearers of the white hats. After that, with the problems inherent in the Atomic Age, the Cold War, and Vietnam, we have moved generously away from that position. With the government distrust engendered in all of those mentioned, a hero tended to be an antihero. This is the person who rises out of nowhere, does a good deed or two for reasons unknown to us, and then disappears (think Mad Max).

One other interesting idea is that perhaps our "antihero" comes from our near-romantic involvement with technology. A computer does not currently come equipped with a moral compass. A machine cannot be said to be good or evil. Perhaps we are embracing technology as a means of avoiding such distinctions.

Does a society need to have heroes or monsters? Joseph Campbell would most likely say yes to heroes. And there is an argument for monsters as well. Both can teach us how to act in different situations.

With the "new" antihero, no longer is good or moral behavior attached to the positive deed done. There are no clear reasons for that, but the idea can possibly be defined and clarified by looking at the rise of technology. Currently, we want to believe that our machines cannot think (although there is an argument on that topic alone). A machine can do things a human cannot; the machine is not motivated by, or controlled by, concepts of good or evil.

We install a great deal of faith in our machines. We are seriously addicted to them. Thus we can "admire" a machine that does not need such old-fashioned constructs like morality. Some people believe that science will end our fascination with monsters. As we learn more, we are less superstitious. We know why the sun comes up and why eclipses occur, but it is important to remember that our fascination with Freud's uncanny carries over into our feelings about machines and technology.

---

[6] Paul Hiebert, "Are All Movie Heroes the Same Person?" Pacific Standard, last modified June 20, 2014, http://www.psmag.com/books-and-culture/movie-heroes-person-joseph-campbell-monomyth-83796

DISCUSSION QUESTIONS

1. Do you think that the rise of technology might have caused the decline in good versus evil distinctions? Why or why not?

2. If a myth is something that we, as a people, believe in, what myths do we believe in today and why?

3. Who are your heroes? Who are the heroes of our current society? Do you think that has changed over the years? Why or why not?

4. Define a monster. If God is that which is most important—then where do monsters fit in? /Are they "evil"? How does God allow that to happen? Why do we look for Bigfoot?

5. Do all societies have monsters? Why or why not? Is a monster made or born? If we revere technology and think it will answer problems, will we develop a "monster" that is a system or a piece of technology?

6. What purpose, if any, do they serve? Should we kill a monster or rehabilitate it?

7. Do you think we NEED monsters? Should children see how monsters behave?

8. Do you think monsters help us to develop ethical considerations or moral behavior? Like a "false" challenge that is not really threatening ("what would you do if. . . .") Do you imagine how you would have behaved given a 911 or an attack on a group of people—student shootings, whatever. Is that a good thing?

9. How about the future? Are cyborgs our new fear? Do you think our "monsters" have become disembodied?

LITERARY SOURCES—FICTION

STRUWWELPETER; MERRY STORIES AND FUNNY PICTURES

Heinrich Hoffman, available online: http://www.gutenberg.org/files/12116/12116-h/12116-h.htm

GRIMM'S FAIRY STORIES

Colored Illustrations by JOHN B. GRUELLE

Pen and Ink Sketches by R. EMMETT OWEN

Available online: http://www.gutenberg.org/files/11027/11027-h/11027-h.htm

Hans Andersen's Fairy Tales, First Series

Edited by J. H. Stickney, Illustrated by Edna F. Hart

Available online at: http://www.gutenberg.org/files/32571/32571-h/32571-h.htm

RESOURCES AND REFERENCES

The Reverend J. A. L. Singh, The Diary of the Wolf-Children of Midnapore http://www.midnapore.in/wolf-children-of-midnapore/wolf-children-of-midnapore8.html

Wolfgang Saxon, NY Times, Joseph Campbell, Writer Known For His Scholarship on Mythology, 11/2/87 http://www.nytimes.com/1987/11/02/obituaries/joseph-campbell-writer-known-for-his-scholarship-on-mythology.html)

In Search of Myths and Heroes, PBS http://www.pbs.org/mythsandheroes/teachersguide.html

Paul A. Trout, Why we invented monsters; How our primate ancestors shaped our obsession with terrifying creatures. Salon, 12/3/11 http://www.salon.com/2011/12/03/the_evolution_of_monsters/

Robert Roy Britt, Monsters, Ghosts and Gods: Why We Believe, Live Science, 8/18/2008 http://www.livescience.com/5046-monsters-ghosts-gods.html

Christopher H. Whittle, Development of Beliefs in Paranormal and Supernatural Phenomena, Skeptical Inquirer, Vol. 28.2, March/April 2004 http://www.csicop.org/si/show/development_of_beliefs_in_paranormal_and_supernatural_phenomena

Videos about Children Raised in the Wild http://www.animalplanet.com/tv-shows/raised-wild/

Joseph Campbell and the Power of Myth with Bill Moyers | Star Wars | PBS https://www.youtube.com/watch?v=2F7Wwew8X4Y

Social Issues: Embedded Technology, The Future of Humanness

"Technology is nothing. What's important is that you have a faith in people, that they're basically good and smart, and if you give them tools, they'll do wonderful things with them."

—Steve Jobs

The twentieth century has brought us a "smart" phone that includes a number of devices that once would have been too cumbersome to carry around. Each generation has continued the concept of "smallification." The traditional desktop at the turn of the century had more computing power, generally speaking than the room-sized first computers from the mid-twentieth century. We can listen to music (replacing various music player types), communicate (replacing our phones), search for information online (replacing desktop computers and, in one sense, libraries), and many other functions all with a fairly small device held in our hands.

The rapid availability of information on such a device is virtually commonplace now. However, it can be both an economic and a generational device. Smart phones might be ubiquitous, but they cost money. At this point, access also costs money. The concept of "free" information is still a thing of the future. They are also small, which can cause problems with the current generation of older people. Small devices have small controls and small visual display screens. At this point in time, eyesight tends to fade with age. While medical discoveries might prevent or stop that problem, it is an issue. Few people over 60 are willing to watch anything on a small screen, for example.

One possible future is the idea or embedding technology within your body. This would allow near constant access. Scientists are working on the "how" of producing energy for such a device to work. As batteries get smaller, we are also looking into using the body's own energy sources to power such a device. On the face of it, this seems like a good idea. If you had access to, say, a map—visible only in your own eye—you would be better able to navigate and find your way. So far, more primitive versions of this like Google Glass™ have not been overwhelmingly successful. The "eyeglass" concept is a bit obvious and difficult to assimilate. Perhaps people do not want to spend time with someone wearing such a device; perhaps it is the idea of the possibility of "big brother" watching you that causes such disquiet.

Concept of uploading data to businessman
Shutterstock/Patrick Foto

There are two obvious sides to this issue. One is the problem with the social construct issue. Walking alone with someone who has more information than you do can be helpful but also divisive. However, such technology could be invaluable to someone who cannot see (or hear, for that matter). In the medical field, it could be invaluable to have implanted medical information within your body. Soldiers, today, for example, can have implanted chips with their medical history on them. This could save their lives in a time of rapid triage. It also could solve the problem of the obvious visual difference in appearance. If you have a physical or medical disability, such a device might eliminate your feeling of "being different."

One of the "Terminator"-type scenarios in science fiction is the idea of a data upload (or download). If we could take what we know (in any subject or format) and upload it to a computer, this could be very helpful if we had some sort of brain damage or dementia in the future. The ability for someone (other than you) to control an upload or a download of that information is a possible nightmare waiting to happen. Lawyers are standing by for all those probable lawsuits.

The ramifications of such embedded technology would apply to almost everything. If you could access a training video on how to accomplish a specific technical task instantly in the field, this would remove the need for a book or any handheld device. Such technology might also be considered imminently divisive. There would be an economic divide; poor people probably would be the last to have access. And there is also the issue of determining exactly what makes us "human." If we continue to constantly "upgrade" our capabilities, would this be evolution or the decline of humanity as we know it (or both)?

The military has its own division of research into these types of devices. DARPA (The Defense Advanced Research Projects Agency) is funding research into battlefield applications of embedded technology:

"In developing a flexible, miniaturized synthesis and manufacturing platform, Battlefield Medicine will leverage continuous flow approaches that will, if successful, pave the path forward for enabling distributed, on-demand medicine manufacturing capabilities in battlefield and other austere environments. Additionally, the platform would have built-in flexibility to produce multiple types of therapeutics through its modular reaction design. The ultimate vision for Battlefield Medicine is to enable effective small-batch pharmaceutical production that obviates the need for individual drug stockpiling, cold storage, and complex logistics."[1]

Sounds good, if a bit verbose. What does it actually mean? You can read this to say that DARPA will be developing medical applications that can be applied "on the fly" whenever needed. In a military situation, this could be saving lives. However, one of the problems with embedding technology in your body is that there is always the possibility of being controlled by such technology. If the technology, for example, is monitoring your body systems, does that mean that some "controller" somewhere could decide that you were no longer "capable" and act accordingly?

When we begin to rely on technology, what are we giving up? We are obviously gaining better technical accuracy, but we are possibly losing control. Today, for example, many schools are actively considering abandoning the idea of teaching cursive writing. The concept behind this is that we no longer "need" such a thing. Almost

---

[1] The Defense Advanced Research Projects Agency, "Battlefield Medicine," accessed March 29, 2015, http://www.darpa.mil/Our_Work/BTO/Programs/Battlefield_Medicine.aspx.

everyone is now communicating via technical devices of some sort. Here is the problem, however. If you write someone a note, at the point of receipt, there are only two people who know the contents of that note in most cases. Obviously, someone could be looking over your shoulder, but the point remains that there is no other "trail" of information. If you type a note on any device, there is almost always a trail. This can be seen as a good thing, but it can also be considered a dangerous thing. A trail of information on a technical device is rarely ever a one-directional source of communication.

© Shutterstock/Juergen Faelchle

DISCUSSION QUESTIONS

1. Would you have electronics embedded in your body? Why or why not?
2. What kind of technology do you think should or could be effectively put into your body?
3. Do you think that we will evolve along with the technology (so that age or other physical factors will not have an impact)?
4. Do you see this as a step in evolution? Or simply a mechanical method of using new technology?
5. Do you think that the religious factions will oppose or agree? Do you see a difference with different religions?
6. What kind of scenario might cause people to distrust this technology?

RESOURCES AND REFERENCES

Gage, Greg, Ted Talk, "How to control someone else's arm with your brain." 3/2015 http://www.ted.com/talks/greg_gage_how_to_control_someone_else_s_arm_with_your_brain?language=en

Belfiore, Michael, "Embedded Technologies: Power From the People. 8/2010 http://www.smithsonianmag.com/40th-anniversary/embedded-technologies-power-from-the-people-1090564/?no-ist

McRae, Lucy, Ted Talk, "How can technology transform the human body?" 2/2012 https://www.ted.com/talks/lucy_mcrae_how_can_technology_transform_the_human_body?language=en

Fyord, "Why the Human Body Will Be The Next Computer Interface." 3/5/2013 http://www.fastcodesign.com/1671960/why-the-human-body-will-be-the-next-computer-interface

Pepitone, Julianne, "Cyborgs Among Us: Human "Biohackers" Embed Chips In Their Bodies." 7/11/2014 http://www.nbcnews.com/tech/innovation/cyborgs-among-us-human-biohackers-embed-chips-their-bodies-n150756

GENDER ISSUES

In a 1993 issue of the New Yorker magazine, there was a famous cartoon by Peter Steiner of a dog at the computer. The caption read: "On the Internet, nobody knows you're a dog."

Perhaps, with the advent of technology, we will cease to ascribe certain roles and behaviors based on gender. If you teach in an online class, it is possible to spend an entire semester without being absolutely certain what gender a student might be. Which begs the question of whether such classification actually matters. The dark side of this coin, however, is that technology can also be used to shamelessly bully and vilify someone without the remotest degree of responsibility or culpability. Your identity can be stolen or your belief structure questioned without you even being aware of it.

While the social construct is moving ever closer to a more gentle and accepting idea about the importance of gender, technology has made it easier and easier to sling insults and hide. The Internet troll exists throughout any computer activity; someone can attack you and you have no possibility of defending or responding without escalating the attack.

"On the Internet, nobody knows you're a dog."
Peter Steiner The New Yorker Collection/The Cartoon Bank

In 2013, game developer Zoe Quinn released the video game *Depression Quest* via an online platform. Since video gaming has been primarily a male enclave since its origins, some people decided to attack her online. They began with a practice known as "doxing" (finding out personal information and releasing it online) and escalated from there into a series of vicious personal attacks including threatening rape and physical harm. It was an organized campaign of hate directed against women. It also provided us with a good picture of the dark side of technology. If you can attack someone anonymously and incite

Shutterstock/thodonal88

others to actual violence, then perhaps new laws and new methods of enforcement need to occur.[1]

Technology can be good or bad depending on its application. Our current technology, for example, can actually allow you to pick and choose the gender of your future child. In some societies, this has been used to an extent that gender balance in a society is severely skewed. Before the advent of new technology, the ratio of males to females was pretty consistent throughout the world. Generally speaking, there are about 103–107 males born for every 100 females.[2] Now that we can safely predict the sex of a baby before birth, there has been a steady rise in gender-based abortion—particularly in countries with a preference for male children. China, for example, implemented a one-child policy, which made the situation worse. By the 1990s, India, South Korea, and China all had too many boy babies. Today, the end result of those decisions is that there are fewer females of child-bearing age in those countries.[3] In India, there have been recent studies that imply that the violence done to women could be related to the excess number of males born.[4]

The unfortunate by-product of the negativity surrounding this science is that pre-implantation genetic diagnosis (PGD) can also be used to see genetic abnormali-

Shutterstock/Sanzhar Murzin

ties. While the religious right might see such a thing as encouraging abortion, the reality of it is that such science can help us immeasurably in the future prevention of abnormalities as well. If we decide not to do the testing because it is "unethical" to consider terminating a pregnancy, we are also losing invaluable information that might someday prevent such problems. For example, if you know that you carry the possibility of a dire genetic disease, you could perhaps have your embryos checked beforehand—and only fertilize and plant an embryo that was disease free.[5]

Technology also has the capability to reduce the differences between genders. Imagine if

[1] Caitlin Dewey, "The Only Guide to Gamergate You Will Ever Need to Read," *The Washington Post*, October 14, 2014, http://www.washingtonpost.com/news/the-intersect/wp/2014/10/14/the-only-guide-to-gamergate-you-will-ever-need-to-read/.

[2] Therese Hesketh and Jiang Min Min, "The Effects of Artificial Gender Imbalance," *EMBO Reports* 13, No. 6 (June 2012), 487–492, doi: 10.1038/embor.2012.62.

[3] Ibid.

[4] Anjani Trivedi and Heather Timmons, "India's Man Problem," *The New York Times*, January 16, 2013, http://india.blogs.nytimes.com/2013/01/16/indias-man-problem/?_r=0.

[5] Gina Kolata, "Ethics Questions Arise as Genetic Testing of Embryos Increases," *The New York Times* (February 3, 2014), http://www.nytimes.com/2014/02/04/health/ethics-questions-arise-as-genetic-testing-of-embryos-increases.html.

every member of a military force had an exoskeleton. It is possible that women who might be less physically strong might be able to equalize the sex-linked differences.

Modern media has generally done a fairly poor job of gender equality. The Bechdel test (named after the American cartoonist Alison Bechdel) was a method of measuring gender equality in films and fiction. The test is very simple (and thus a bit flawed): The media has to have at least two women in it. Those two women have to actually talk to each other, and they cannot be talking about a guy.[6] It is surprising how many movies fail this test. Until our culture can accept the concept that neither sex is "better than" the other, we will continue to struggle with issues of gender.

Another issue is the problem of gender identity. We once believed that the gender that you were born with was always the correct one. Some people still believe that homosexuality is a choice and that it has no actual scientific basis. Those same people, of course, do not consider heterosexuality to be any kind of choice, but rather the acceptable norm.

In addition, any time science tries to label or effectively measure the sex of something, it is always problematic. The International Olympic Committee, for example, has been trying to figure out how to tell if someone is female or male for some time. They base their decisions on testosterone levels, but no one knows just what those levels might be. It is also possible to raise or lower the levels medically.[7]

In an article in Nature, Claire Ainsworth reported that research is beginning to imply that our current notion of two sexes may be incorrect and that, in fact, there may be a much wider spectrum of sex than we had previously supposed.[8] There have been documented cases of people who have spent their entire lives believing that they were one sex (had children, lived as one sex) only to discover (usually be accident) that they also carried a set of chromosomes for the opposite sex as well. We don't know why; some theories include the idea of twin embryos might have merged at a critical point in gestation. As the author states:

"That the two sexes are physically different is obvious, but at the start of life, it is not. Five weeks into development, a human embryo has the potential to form both male and female anatomy."[9]

Scientists now believe that the shift from one sex to another might also be happening in adults. They have managed to deactivate certain genes in adult mice that might change sexual development at any point. This is a biological approach to the idea of homosexuality being something that occurs at birth versus some sort of learned behavior.

Unfortunately, some psychiatrists and other therapists did believe at one point that therapy could change a homosexual into a heterosexual. This therapy called conversion therapy or reparative therapy had quite a following. It still exists in some form today, especially with those who follow religious tenets that imply that heterosexual behavior is normal while homosexual behavior is not. There are few (if

---

[6] "Bechdel Test Movie List," accessed March 29, 2015, http://bechdeltest.com.

[7] "IOC Regulations on Female Hyperandrogenism," June 22, 2012, http://www.olympic.org/Documents/Commissions_PDFfiles/Medical_commission/2012-06-22-IOC-Regulations-on-Female-Hyperandrogenism-eng.pdf.

[8] Claire Ainsworth, "Sex Redefined," Nature 518, no. 7539 (2015): 288–291, http://www.nature.com/news/sex-redefined-1.16943.

[9] Ibid.

Shutterstock/Dziobek

any) true "success stories" with this kind of therapy. Most were not only ineffective, but also extremely dangerous.

Many studies were done in order to prove this concept. In one, done by Robert L. Spitzer in 2003, he claimed that 200 people in the study had been "changed" by the therapy. After being homosexual, they then reported heterosexual ideation and beliefs.[10] In 2012, he apologized for the "deeply flawed study" and admitted that this type of therapy was not beneficial.[11] As he himself stated, it would be difficult, if not impossible, to verify any self-reported behavior by simply asking.

As we moved into the twenty-first century, the concept of tolerance for sexual differences has hopefully moved forward into acceptance. Gay marriage is now legal in many states, and gay adoption is also becoming more commonplace. Some people still reject the idea that homosexuality is not a matter of choice, but ongoing education efforts may help to change that in the future. Today, the idea of "gay rights" is slowly being expanded to include other "categories." The American Civil Liberties Union formed its Lesbian Gay Bisexual & Transgender Project as an outgrowth of their first case in 1936.[12] Today, the original LGBT rights have been expanded to include many more types of gender differences.

Science and technology can provide the underpinnings for our understanding of sexual preferences. However, society moves slowly when it comes to changes like this and some religions can slow the change even more.

In her mind-bending science fiction work, "The Left Hand of Darkness," Ursula K. Le Guin envisioned a society where the "acceptable" people in a society were all basically hermaphrodites (having the characteristics of either sex). A person could take on male or female characteristics depending on the situation or context. Perhaps we might someday learn to use this idea to help us realize that sex and gender are not always the concrete concepts that we currently believe.

---

[10] Robert L. Spitzer, "Can Some Gay Men and Lesbians Change Their Sexual Orientation? 200 Participants Reporting a Change from Homosexual to Heterosexual Orientation," *Archives of Sexual Behavior* 32, no. 5 (October 2003): 403–417, http://www.jpsych.com/pdfs/Spitzer,%202003.pdf.

[11] B. A. Robinson, "An Analysis of Dr. Spitzer's 2001 Study Into Whether Adults can Change Their Sexual Orientation," *Ontario Consultants on Religious Tolerance*, last modified on October 6, 2012, http://www.religioustolerance.org/hom_spit5.htm.

[12] "LGBT Rights: Lesbian Gay Bisexual & Transgender Project," American Civil Liberties Union, accessed March 29, 2015, https://www.aclu.org/lgbt-rights.

DISCUSSION QUESTIONS

1. If you had a 4- or 5-year-old daughter, would you rather she played with a fashion doll or a female action figure? Defend your choice.

2. Do you see any difference in heroic behavior between different sexes? How?

3. Would you look for something different in female leadership than you would in male? Why?

4. Do you think gender is strictly biological? Or do you think other factors might be involved (nature vs. nurture).

5. Do you think we need new and/or better laws to prevent cyberbullying? What kind of laws should we have and who should decide?

6. What is your opinion of the Bechdel Test? Do you think we should be monitoring our media for gender equality? If so, how? If not, why?

RESOURCES AND REFERENCES

Sidhu, Jasmeet, How to buy a daughter. Slate, 9/14/12, http://www.slate.com/articles/health_and_science/medical_examiner/2012/09/sex_selection_in_babies_through_pgd_americans_are_paying_to_have_daughters_rather_than_sons_.html.

Kolata, Gina, Ethics Questions Arise as Genetic Testing of Embryos Increases, NYT, 2/3/2014, http://www.nytimes.com/2014/02/04/health/ethics-questions-arise-as-genetic-testing-of-embryos-increases.html.

International Olympic Committee, IOC Regulations on Female Hyperandrogenism, 2012, http://www.olympic.org/Documents/Commissions_PDFfiles/Medical_commission/2012-06-22-IOC-Regulations-on-Female-Hyperandrogenism-eng.pdf.

Kaur, Ravinder, Mapping the Adverse Consequences of Sex Selection and Gender Imbalance in India and China, Economic & Political Weekly, 8/31/2013, vol. xlviii no 35, http://www.academia.edu/4354147/Mapping_the_Adverse_Consequences_of_Sex_Selection_and_Gender_Imbalance_in_India_and_China.

Trivedi, Anjani and Timmons, Heather, India's Man Problem, NY Times, 1/16/13, http://india.blogs.nytimes.com/2013/01/16/indias-man-problem/?_r=0.

Dewey, Caitlin, The only guide to Gamergate you will ever need to read. Washington Post, 10/14/2014. Retrieved 3/12/2015, http://www.washingtonpost.com/news/the-intersect/wp/2014/10/14/the-only-guide-to-gamergate-you-will-ever-need-to-read/.

Khaw, Cassandra, How a Gamergate target is fighting online harassment, Engadget, 2/23/2015, retrieved 3/12/2015, http://www.engadget.com/2015/02/23/crash-override-interview/.

Wagner, Kyle, The future of the culture wars is here, and it's Gamergate. Deadspin, 10/14/14, retrieved 3/12/15, http://deadspin.com/the-future-of-the-culture-wars-is-here-and-its-gamerga-1646145844.

Gender inequality in film, New York Film Academy Blog, https://www.nyfa.edu/film-school-blog/gender-inequality-in-film/.

Bechdel Test Movie List, http://bechdeltest.com.

Smith, Stacy L. and Cook, Chrystal Allene, Gender Stereotypes: An analysis of popular films and TV, Conference 2008, http://seejane.org/wp-content/uploads/GDIGM_Gender_Stereotypes.pdf.

National Geographic explains the biology of homosexuality https://www.youtube.com/watch?v=saO_RFWWVVA.

The Sissy Boy Project, CNN 360

    Part 1: https://www.youtube.com/watch?v=A-irAToviFo

    Part 2: https://www.youtube.com/watch?v=t_AeCUSXfmU

    Part 3: https://www.youtube.com/watch?v=rI6-g5hnRmg

Depression Quest (Video Game) http://www.depressionquest.com

Social Issues: Media and Communication

"You furnish the pictures and I'll furnish the war."

—William Randolph Hearst

In 1897, William Randolph Hearst, the owner of the New York Journal, sent the artist Frederic Remington to Cuba. He was to provide sketches of the situation in Cuba at that time. Remington spent several days there, making sketches of the alleged "rebellion" that was occurring at the time. U. S. Newspapers were very interested in the situation.

The story is that Remington felt that there was nothing going on. In his cable to Hearst, he stated: "Everything is quiet. There will be no war. I wish to return." The well-known story after that is that William Randolph Hearst sent back a telegram that said, "Please remain. You furnish the pictures, and I'll furnish the war." People have been arguing about it ever since; historians believed that the "yellow journalism" of the times was a contributing factor to the occurrence of the Spanish American War.

There is mixed evidence that the cable from Hearst to Remington ever existed. Hearst claimed that he did not send it; and the cable itself was never found. There are two important lessons here.

1. What we read as "true" may not be true.
2. A picture may convince you when words cannot.

A third important caveat is that, quite simply, this is the digital age. You have no doubt seen those "amazing" pictures of the guy in a wetsuit being hauled into a helicopter as a shark tries to grab him. This is a good example of just how easy it is to manipulate an image.

A picture can have a hefty amount of influence when we are trying to convince someone. In his blog, Michael Moore argued that we should see the bodies of the children at the Newtown massacre. If we actually could see the damage done to a human body by a round of ammunition, it would galvanize people toward gun control.[1] During World War II, no pictures of dead Americans were allowed in the press until a Buna Beach picture taken by George Strock was published in September of 1943.[2]

---

[1] Michael Moore, "America, You Must Not Look Away (How to Finish Off the NRA)," *Mike's Tumblr*, Last Modified March 13, 2013, http://blog.michaelmoore.com/post/85700192513/america-you-must-not-look-away-how-to-finish-off.

[2] Ben Cosgrove, "The Photo That Won World War II: 'Dead Americans at Buna Beach,' 1943," *LIFE*, October 31, 2014, http://time.com/3524493/the-photo-that-won-world-war-ii-dead-americans-at-buna-beach-1943/.

Why does that matter? World War II had been going on for several years at that point, but no pictures of dead Americans had been shown. Seeing those bodies had a much greater impact than simply reading about the number of deaths.

Michael Moore speaks of seeing the body of Emmett Till, a 14-year-old child killed by racists in 1955 in Mississippi. His mother asked that the body be displayed so that people could see the kind of savagery that was done. It was a galvanizing act. If you simply read about such a story in the newspaper, you might have some sympathy but a picture cannot be unseen. While we can argue that human decency should prevent seeing such a picture, its impact cannot be denied.

In 2013, there was a carefully photographed series that documented domestic abuse. The photographer, Sara Lewkowicz, published a series of pictures that clearly displayed the horrors of domestic abuse. There was a firestorm of criticism about the situation; many people felt that, instead of documenting what happened, she should have made more effort to prevent or stop it. The question remains, which had the greater impact? Humans are visual. Pictures have power. If you had simply read the story, you might have dismissed it ("these things happen to those people"). Seeing it makes it far more visceral. The impression can be far greater.[3]

In addition to the aspect of photography, we also have the changing aspect of news reporting today. The idea of "television news" has changed drastically in the last 50 years. In 1949, the Federal Communications Commission (FCC) introduced the policy known as the Fairness Doctrine. This policy required the people who purveyed the news to present it in a "fair and balanced" approach.[4] The idea was that if you wished to obtain a license to present news broadcasting to the American public, then you would be required to provide information on the various sides of the issue. It was designed to protect the public interest. Broadcast media was required by law to provide contrasting opinions of any controversial news items presented. Additionally, there were specific rules involved concerning a personal attack; the person had to be notified and given a chance to rebut the attack within a specific period of time.[5]

This ruling allowed at least the appearance of equal presentation of controversial issues within the television medium. However, the original ruling came under attack. Some journalists believed that requiring this "fair and balanced" approach violated First Amendment rules concerning freedom of speech. In *Red Lion Broadcasting Co., v. Federal Communications Commission* (1969), the Supreme Court had upheld the constitutionality of the doctrine, ruling that the FCC did have the congressionally delegated authority to enact such rulings. In that case, a politician had been maligned on the air and he had sought equal time to argue his side. It was denied.

However, by the 1980s, the FCC began to re-examine the Fairness Doctrine. Opinions then seemed to be based upon not only the free speech argument, but also the idea of marketplace conditions. In addition, the idea that such regulations might prevent the airing of certain materials (due to the cost of providing both sides and/or

---

[3] Sara Naomi Lewkowicz, "Shane and Maggie," Accessed April 30, 2015, http://www.saranaomiphoto.com/Shane-and-Maggie/1/.

[4] Federal Communications Commission Reports, Appendix, Docket No. 8516 http://transition.fcc.gov/ftp/Bureaus/Mass_Media/Databases/documents_collection/490608.pdf

[5] Kathleen Ann Ruane, "Fairness Doctrine History and Constitutional Issues, Congressional Research Service, 7/13/2011, Accessed April 30, 2015, http://fas.org/sgp/crs/misc/R40009.pdf.

possible legal challenges) might tend to prevent public access to information. A radio station might cancel a program entirely if it was too expensive to provide information from more than one source. Since the Fairness Doctrine also required the government to decide—up-front—just what might be in the public interest, it also was subject to concerns of excess government intervention. By then, there were a great many more available radio and TV stations available as well. The Fairness Doctrine was repealed in 1987.

From a freedom of speech standpoint, this was a valid decision. However, from that point forward, it became more and more possible for a TV news outlet to slant the news in any way the owners or journalists wished. One can argue that such a trend might have given us the extreme polarity seen in the news today. While many stations continue to attempt to provide fair and equitable coverage, some do not. People who watch television news tend to have favorites that support their own views. Thus, the idea of seeing both sides of an issue could easily fall by the wayside. In addition, if a news station or TV broadcaster leads the evening news with X story, it tends to imply that the subject and/or news item is of critical importance, which can also sway the public opinion. According to Dylan Matthews of the Washington Post, "The decision has been *credited* with the explosion of conservative talk radio in the late '1980s and early '1990s"[6]

According to a study done by the Pew Research Center in 2013:

"In the evening, Fox News boasts a lineup of conservative talk show hosts while MSNBC features a team of liberal ones. CNN, the original cable news outlet, has built its brand around national and global reporting of breaking news events. It also airs opinion in prime time, but includes commentators from both the right and the left."[7]

In a Pew Research Center report in 2014, consistent liberals used a variety of news sources like CNN, MSNBC, NPR, and NYT to learn about the news. In contrast, 47% of consistent conservatives utilized Fox News only. Liberals tended to trust news sources, while conservatives tended to distrust most news services:

"Overall, the study finds that consistent conservatives:

> Are tightly clustered around a single news source, far more than any other group in the survey, with 47% citing Fox News as their main source for news about government and politics.

> Express greater distrust than trust of 24 of the 36 news sources measured in the survey. At the same time, fully 88% of consistent conservatives trust Fox News.

> Are, when on Facebook, more likely than those in other ideological groups to hear political opinions that are in line with their own views.

> Are more likely to have friends who share their own political views. Two-thirds (66%) say most of their close friends share their views on government and politics.

---

[6] Dylan Matthews, "Everything You Need To Know About The Fairness Doctrine In One Post," Washington Post, 8/23/2011 http://www.washingtonpost.com/blogs/wonkblog/post/everything-you-need-to-know-about-the-fairness-doctrine-in-one-post/2011/08/23/gIQAN8CXZJ_blog.html).

[7] Kenneth Olmstead, et al., Pew Research Center, 10/22/2013 http://www.journalism.org/2013/10/11/how-americans-get-tv-news-at-home/).

By contrast, those with consistently liberal views:

> Are less unified in their media loyalty; they rely on a greater range of news outlets, including some—like NPR and the New York Times—that others use far less.

> Express more trust than distrust of 28 of the 36 news outlets in the survey. NPR, PBS and the BBC are the most trusted news sources for consistent liberals.

> Are more likely than those in other ideological groups to block or 'defriend' someone on a social network—as well as to end a personal friendship—because of politics.

> Are more likely to follow issue-based groups, rather than political parties or candidates, in their Facebook feeds."[8]

Most students would tend to see this as a generational age gap type issue. However, it is also possible that dependence on a given news source can be categorized by race, gender, age, and a variety of other measurements. It has been established that "young people" are generally watching less television than they have in the past. Unfortunately, using the Internet as a news source can be problematic. First, you are generally "cherry-picking" your news and your source. Second, you may not seek out opposing views or see the broader picture when you self-select what you will see.

In his Ted Talk in 2011, Eli Pariser speaks of "filter bubbles" when we seek to do Internet research on any topic. The idea here is that computer search algorithms want (if you can personalize such a thing) to make you happy. If you, for example, search out information about your health or some kind of medical treatment and you always read articles about holistic medicine, then sooner or later, your computer will start moving those up to the early pages of your search. You have effectively trained the machine to follow your desires. Unfortunately, it tends to make you think, perhaps, that many people agree with you ("Look, the first 5 pages are holistic medicine in my search—clearly that's the best way to go.") Having more access to information may not always end up making us smarter.

---

[8] Amy Mitchell, et al, Political Polarization & Media Habits, Pew Research Center, 10/21/2014, http://www.journalism.org/2014/10/21/political-polarization-media-habits/.

DISCUSSION QUESTIONS

1. Do you think seeing a picture is essential to understanding some stories?

2. Are you the kind of person that needs to see an image or are you the kind of person that thinks such things should be banned? Why?

3. Do you think the media—by what it portrays—can affect human belief and behavior?

4. Have you ever bought a product advertised in the media? Why or why not? What influences you to make a purchase?

5. Do you think the media controls you or how you think?

6. How do you access the news?

7. How often do you read ALL of an article on the Internet?

RESOURCES AND REFERENCE ARTICLES

Steinmetz, Katy, The death of the fairness doctrine. Time Magazine, 8/23/2011, retrieved 3/6/15, http://swampland.time.com/2011/08/23/the-death-of-the-fairness-doctrine/.

Hershey, Robert D., Jr., F.C.C. votes down fairness doctrine in a 4-0 decision, NY Times, 8/5/1987 as retrieved on 3/6/2015, http://www.nytimes.com/1987/08/05/arts/fcc-votes-down-fairness-doctrine-in-a-4-0-decision.html.

Blake, Meredith, The big bang of older TV viewers, LA Times, 2/22/14, as retrieved on 3/6/2015, http://www.latimes.com/entertainment/tv/showtracker/la-et-st-aging-tv-audience-20140223-story.html#page=1.

Staff, Marketing Charts, Are Young People Watching Less TV? (Updated – Q3 2014 Edition), as retrieved on 3/6/2015, http://www.marketingcharts.com/television/are-young-people-watching-less-tv-24817/.

Television Watching Statistics, 12/7/13, http://www.statisticbrain.com/television-watching-statistics/.

Pariser, Eli, Ted Talk, March 2011, "Beware Online Filter Bubbles" http://www.ted.com/talks/eli_pariser_beware_online_filter_bubbles

Social Issues: Technology and Personal Relations

> "I fear the day that technology will surpass our human interaction. The world will have a generation of idiots."
>
> —Albert Einstein

As each new generation emerges, they tend to develop their own particular kind of communication. We have gone from the invention of the telephone, to email, to the idea that all-important messages can be transmitted via text. By the time this textbook is printed, that will, no doubt, be replaced with something else.

The most important issue with social media is trying to determine if it is a valid type of communication. Numerous studies are conducted regularly on what the effect of this kind of communication might be. In her Ted.com talk of February, 2012, Sherry Turkle made this comment:

"People text or do email during corporate board meetings. They text and shop and go on Facebook during classes, during presentations, actually during all meetings. People talk to me about the important new skill of making eye contact while you're texting."[1]

On the face of it, people of a certain generation might conclude that this is simply rude. You should "always" pay attention to a speaker and to do otherwise is wrong. This causes a certain generational gap. Students today believe quite firmly that they can multitask; this, to most college students, is "appropriate" behavior. Older people might decide that you were not engaged in the "work" of the classroom.

The "generational" side of this is interesting simply because technology is driving us toward smaller and smaller gadgets (until, possible, we simply embed them). To an older person, a small gadget is problematic for both physical dexterity and visual issues. As we "smallify" the technology, we are already establishing yet another generational divide.

Another issue is whether such technical means of communicating is actually a viable and decent way to contact someone. It works incredibly well for the simple quick contact point—"meet me at x"—but it can be devastating (or moving) in its very simplicity. Think about the boyfriend or girlfriend that breaks up with you via text (or, even worse, via a texted picture of your one true love with someone else).

---

[1] Sherry Turkle, "Connected, but Alone?", *TED Talk Subtitles and Transcript*, April 2012, http://www.ted.com/talks/sherry_turkle_alone_together/transcript?language=en.

As Sherry Turkle put it:

"Why does this matter? It matters to me because I think we're setting ourselves up for trouble—trouble certainly in how we relate to each other, but also trouble in how we relate to ourselves and our capacity for self-reflection. We're getting used to a new way of being alone together."[2]

It is very difficult to sort this out without the older generation being told that they (as usual) no longer "get it." But one issue definitely remains. What, if any, is the problem with multitasking? The evidence is still out on this as well. Some studies have implied that the rise of techno-gadgets has driven the number of ADHD diagnoses up.[3,4]

Some people assume that their children could not have ADHD because they can concentrate for hours on a video screen. However, as Dr. Perri Klass points out:

"In fact, a child's ability to stay focused on a screen, though not anywhere else, is actually characteristic of attention deficit hyperactivity disorder."[5]

One interesting aspect of this is that video games can also be used as a form of biofeedback to assist children with ADHD to become more focused on their work.[6]

There is also some research that implies that the ability to multitask may be a genetic issue. Studies have shown that a small percentage of high-school-aged students can actually perform better with a variety of media inputs at the same time. Whether this is true currently, we also have no way of knowing whether the use of the technology drove the behavior or vice versa.[7]

Another argument has to do with the development of creativity. Some English teachers believe that the rise of "short" communication devices like texting or Twitter will have a negative impact on your ability to write. It is an interesting point, but the opposite aspect is equally possible. If we simply want to impart information quickly, then social media has a valuable place. Sometimes our very verbosity tends to impact the knowledge we need. The best example of this would be the insert in drug packaging. Hardly anyone reads it because it is mind-numbingly verbose. Great literature, however, is a different issue.

There are also several theories that imply that our brains are simply being rewired. Someone who has been immersed in gadgetry since birth might have the ability to simply rewire their brains to meet the demands of the new technology.

---

[2] Ibid.

[3] Margaret Rock, "A Nation of Kids with Gadgets and ADHD: Is technology to blame for the rise of behavioral disorders?" *Time Magazine*, July 8, 2013, http://techland.time.com/2013/07/08/a-nation-of-kids-with-gadgets-and-adhd/print/.

[4] Larry Rosen, "ADHD and Technology: Helping Our Children Reclaim Their Focus and Attention," *Huffington Post* (blog), November 5, 2014 (5:59 am), http://www.huffingtonpost.com/dr-larry-rosen/adhd-and-technology-helpi_b_6096168.html.

[5] Perri Klass, "Fixated by Screens, but Seemingly Nothing Else," *The New York Times* (New York, NY), May 9, 2011, http://www.nytimes.com/2011/05/10/health/views/10klass.html?_r=2.

[6] "NASA Research Study With ADHD Children," *SmartBrain Technologies*, Accessed March 29, 2015, http://www.smartbraintech.com/nasa.asp.

[7] The Advisory Board Company, "For about 15% of teens, media multitasking may boost efficiency, Focus on teens finds some people perform better when distracted," *The Daily Briefing*, October 16, 2014, http://www.advisory.com/daily-briefing/2014/10/16/for-about-15-of-teens-media-multitasking-may-boost-efficiency.

"Over the course of a lifetime, a person 'trains' his neural network to be as efficient as possible at responding to stimuli in his own environment.

The array of stimuli in the modern world—think email, flashing web pages, interactive video games—is distinctly different than anything the brain has had to adapt to in the history of human evolution.

As a result, our brains are being forced to adapt to a changing at an astronomical pace."[8]

In a study done by the Kaiser Family Foundation in 2010, the amount of time spent by children with technology was considered problematic.

"8-18-year-olds devote an average of 7 hours and 38 minutes (7:38) to using entertainment media across a typical day (more than 53 hours a week). And because they spend so much of that time 'media multitasking' (using more than one medium at a time), they actually manage to pack a total of 10 hours and 45 minutes (10:45) worth of media content into those 7."[9]

The question remains: just how "social" is social media and technology? We tend to look at our computers as essential parts of our lives.

There have been a number of studies on the subject known as the "CASA Paradigm." CASA, in this instance, stands for Computers Are Social Actors. In a study done by Clifford Nass and others at Stanford University in the early 90s, it was established that people would almost routinely ascribe "human" characteristics to their computers. The research was done by taking certain social science theories on subjects like politeness or gender and simply replacing the word "person" with the word computer.

For example, if you did research on whether people would be more polite if they knew the audience was mostly one gender, then you would change the research to ask if a computer response was given a male or a female voice, did it change the perception of the listener.

"Subjects responded consistently with a belief that a different computer with a different voice is a distinct social actor."[10]

In other words, a computer giving you information in one voice will be interpreted differently if the computer gives you the same information in a different voice.

That would make us personifying our computer. In later chapters, we will discuss the effect of artificial intelligence on our perceptions. If you have trouble accepting the idea of a personal relationship with a computer, then imagine how you would feel if you suddenly lost everything on your hard drive—or if you lost your cell phone. You tend to have an emotional response if either one happens. There have also been studies that showed that people would rate their own beloved computers

---

[8] "How is Digital Technology Changing the Way Kids' Brains Learn?", New Media and Development Communication, accessed March 29, 2015, http://www.columbia.edu/itc/sipa/nelson/newmediadev08/The%20Neuroscience%20of%20Learning.html.

[9] Victoria J. Rideout, Ulla G. Foehr, and Donald F. Roberts, "Generation M2: Media in the Lives of 8- to 18-Year-Olds," *The Henry J. Kaiser Family Foundation*, January 20, 2010, https://kaiserfamilyfoundation.files.wordpress.com/2013/04/8010.pdf.

[10] Clifford Nass, Jonathan Steuer, and Ellen R. Tauber, "Computers are Social Actors," (presentation, Annual Convention of The Association for Computing Machinery, Boston, MA, April 24-28, 1998), http://www.radford.edu/~sjennings15/CASA.pdf.

higher (almost as though they did not want to "hurt their feelings") in a given test. If the same test was given on a different computer, the responses would be different.[11]

The idea of multitasking is not something that we can ignore; however, it is obviously here to stay. One important issue, as Clifford Nass put it,

"It turns out multitaskers are terrible at every aspect of multitasking. They're terrible at ignoring irrelevant information; they're terrible at keeping information in their head nicely and neatly organized; and they're terrible at switching from one task to another." [12]

This could have serious repercussions if, as many believe, the concept of electronic multitasking is here to stay. Research done by Nass has suggested that there might be a reduction in analytic reasoning.[13] One other issue is that multitaskers are known to be relatively poor at ignoring the irrelevant. If you are trying to write a paper or studying and you have a device that tells you that you have a text or email via a sound, it is pretty difficult to stay on track and ignore that. There is also the issue, per Nass, that multitaskers do not perceive the problem. Multitaskers seem to think they do very well at multitasking.[14]

The argument, however, that electronics might be "dumbing us down" is interesting. Originally, most people's brains were simply not wired to think about multiple things. Nass used the example of a hunter who is fixed on one prey. While said hunter might be paying attention to sounds in the forest, he or she would still be pursuing a single goal. Your brain was designed to relate ideas to each other. Thus, the hunter would automatically connect sounds in the forest to location of prey. Today, if we are inundated with many different forms of diversion, our brain might be having trouble "connecting the dots."

Perhaps, this will be an evolutionary thing. Perhaps our brains will slowly reconfigure to accept and utilize this new norm. The greater problem might be our ability to reason and analyze situations. If our brain is striving to relate subjects that are not related, then we may, in fact, be "dumbing down."

Shutterstock/fitim bushati

Studies are currently being done on whether or not social media like Facebook or Twitter are actually meeting social needs for connection and communication. You can argue that verbosity in old-fashioned letter writing detracted from the message being sent. Or you can argue that writing is a skill that should not be lost. But doing so, you might be avoiding the major question. Is this truly "social" interaction? Facebook allows us to share our daily lives to a level of minutiae that can be painful to others. Or we can share moments of our happy "together" lives that might cause pain to someone whose life is not quite so "together." As a medium of social discourse, it is important in its "instant"

---

[11] Ibid.

[12] Clifford Nass, "Stopping Multitasking is Mostly a Losing Battle," *Frontline*, February 2, 2010, http://www.pbs.org/wgbh/pages/frontline/digitalnation/interviews/nass.html.

[13] Ibid.

[14] Ibid.

sharing capacity; as a method of communication, it possibly lacks true connection. It is possible that you could read of someone being obviously and repeatedly depressed and you might be moved to actually do something for that person, but, as a rule, the very superficiality of such a program tends to give us a more comfortable degree of separation from someone's pain. Do we instantly feel better based on the number of "likes" we get? It is hard to tell.

Shutterstock/Sylvie Bouchard

One other problem that should be mentioned is the fact that social media can be very effective and damaging in attempts to bully people. It can be easily manipulated and instantly widespread and public, which only exacerbates the problem. Teenagers can instantly share viciously wrong information and damage lives with the push of a button. Programs like Snapchat, which "promise" that a photo will disappear, can be even more problematic. Anything that appears on any screen can be captured and used to cause harm.

There is also the concept that all of the information shared in such a public setting is fodder for social research. Market branding, quality control, advertising, and all other aspects of commercial appeal can be measured by the number of times we mention something on Facebook or Twitter or any other social media site. In one sense, we are giving away our information for the privilege of using such a platform.

We use programs like LinkedIn to aid in a job search and networking. Such programs help to build community and connect you to others in the same field. If it is true that it is "who you know," then programs like LinkedIn should help you find the job you want. There is no clear-cut data, so far, that such an idea is true, but networking is important in any field. In recent years, LinkedIn lowered the age limit to join to 13, perhaps putting even more social media into the idea of a future career.

Courtesy of Library of Congress

As to communication media like Twitter, the very immediacy can be extremely important. Of course, telling something important within the severe constraints of a certain number of characters may mean that you learn only the shell and not the substance.

DISCUSSION QUESTIONS

1. Do you think you are a good multitasker? What would other people say about you in this capacity?

2. Does multitasking affect how you learn? Do you believe that having a number of input stimuli is better or worse for learning?

3. Do you think access to information can be an economic question? If so, what effect does this have on us as a society?

4. Have you ever been on a social media platform and seen evidence that someone might be dangerously depressed? What would you do?

5. Do you think that you can convey important information within a social media platform?

6. Have you seen any evidence of the misuse of social media? Do you know anyone who was bullied by its use? How can you address that problem?

RESOURCES AND REFERENCE ARTICLES

"America, You Must Not Look Away (How to Finish Off the NRA), Michael Moore, blog, http://blog.michaelmoore.com/post/85700192513/america-you-must-not-look-away-how-to-finish-off.

Ben Cosgrove, The Photo That Won World War II: 'Dead Americans at Buna Beach,' 1943, Time, 10/27/14, http://time.com/3524493/the-photo-that-won-world-war-ii-dead-americans-at-buna-beach-1943/.

Torie Rose DeGhett, "The War Photo No One Would Publish": When Kenneth Jarecke photographed an Iraqi man burned alive, he thought it would change the way Americans saw the Gulf War. But the media wouldn't run the picture. The Atlantic, 8/8/14, http://www.theatlantic.com/features/archive/2014/08/the-war-photo-no-one-would-publish/375762/.

De Chourdury, Munmun, et al, Predicting Depression via Social Media 2013, Association for the Advancement of Artificial Intelligence (www.aaai.org), http://research-srv.microsoft.com/pubs/192721/icwsm_13.pdf.

Konnikova, Maria, How Facebook Makes Us Unhappy, The New Yorker, 9/10/2013 as retrieved on 3/6/2015, http://www.newyorker.com/tech/elements/how-facebook-makes-us-unhappy.

Bine, Anne-Sophie, Social Media Is Redefining "Depression," The Atlantic, 10/28/2018 as retrieved on 3/6/2015, http://www.theatlantic.com/health/archive/2013/10/social-media-is-redefining-depression/280818/.

The Problem with Multitasking, The Energy Project https://www.youtube.com/watch?v=kpBio_nlLME

Smart Brain Technologies https://www.youtube.com/watch?v=2wNrPhBEoZM

Social Media Revolution 2015 https://www.youtube.com/watch?v=jottDMuLesU

Biology and Genetics: Introduction

"It has become appallingly obvious that our technology has exceeded our humanity."

—Albert Einstein

What makes us fully human or fully alive? This is a question that we have pondered for many years. In the novel Frankenstein, Mary Shelley gave us an early version of what might happen when we try to create life.

Mary Shelley was 19 when Frankenstein was written in 1816 (it was published anonymously two years later). She had a very abnormal (for the times) upbringing. Her mother was the famous feminist Mary Wollstonecraft, who died at Mary's birth. Her father was William Godwin, who was a well-known and famous liberal philosopher (and atheist). He quickly remarried.

Mary was raised without any formal education (fairly normal for the time), but she was allowed to spend time "at table" whenever adults came to visit. In her family, this meant that many of the famous luminaries of the time would come for dinner. She fell in love with the poet Percy B. Shelley when she was 16 (he was married to someone else at the time) and finally ran off with him in 1814. He was not a strong believer in monogamy and thought nothing of offering to share her with his friends.

In the year 1815, the Mount Tambora volcanic eruption in Indonesia was believed to have changed the weather over much of the world. The next year became known as "The Year Without a Summer." For unknown reasons, the weather turned uncharacteristically cold and crops died allegedly due to the eruption of the volcano. It was a time when superstition overruled science among the general population.

Mary and Shelley were vacationing with a number of friends including Lord Byron and Mary's younger sister (who was pregnant by Byron). Since they were stuck indoors, they began telling wild ghost tales to entertain each other. They finally reached a point where they decided to go to bed and whoever wrote the best story would "win." Mary claimed that she dreamed Frankenstein that night. There has been some argument in the past that she did not write it—that perhaps Shelley or Byron did, since Mary had written nothing before and little afterward.

Interesting side note: Byron went on to produce only one legitimate child who grew up to be Ada Lovelace—known as the first person to write a "program" for a computer (the Babbage Machine). Ada never met Byron.

The subtitle of the novel Frankenstein is: "The Modern Prometheus." Any student of mythology would immediately have a clue that this was not going to end well. Prometheus was the person who stole Zeus's fire from the sun and he was subsequently

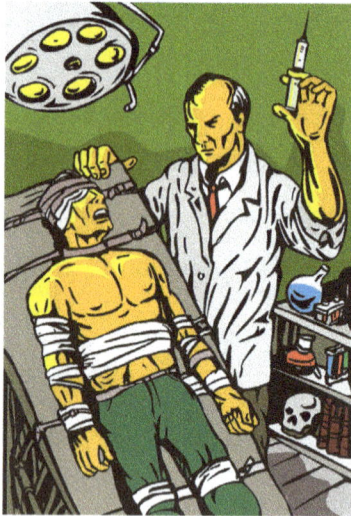

Shutterstock/patrimonio designs ltd

punished by being chained to a rock. Every day an eagle would peck out his liver; by night, the liver grew back to start the cycle again.

In the novel, the main character, Victor Frankenstein, was the doctor who created the monster. By attempting to create life, it could be said that he was attempting to play God. In 1816, this would have been considered horrific in and of itself. It could also be construed as a plot against women (the "natural" procreators). There is also a certain subtext that could be construed as a harsh criticism of religion. The act of hiding what he is doing only makes it worse.

It is a good starting place for a discussion on science and technology and the impact of culture. If you look at what happened in the novel, there are two important points to consider. Which is worse? Creating the monster or abandoning it? You can subtract the religious aspect if you must, but there are many people today who still believe that the act of creating life is not a good idea. The second important issue to consider is do we have an obligation to "protect" what we create? Was Victor wrong for trying to do what he did or was he wrong to abandon it after he realized what could happen?

In another sense, it is also important to realize that Victor began his studies to reanimate the dead after the death of his mother. As with many scientific discoveries, the origin is in good intentions. He was trying to do a good thing. Does this in any way excuse what happened?

In a famous recent study, Australian scientists were attempting to curb the mice population using genetic engineering. The idea was to inject a gene into the mousepox virus that would cause infertility. Instead, it was lethal to the mice (even mice

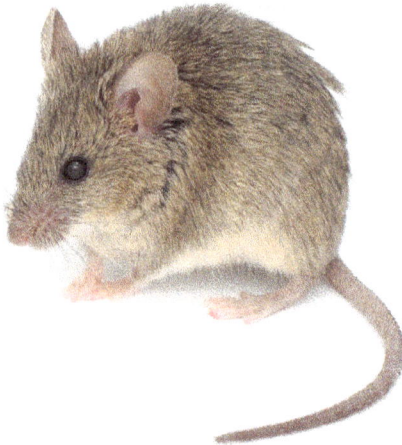

Shutterstock/Tsekhmister

that had been vaccinated against mousepox). It is too short a step to move from an allegedly "harmless" piece of research for a specific solution to surmise that terrorists could use the same techniques to destroy some part of the world. This leads to another problem: should you, as an ethical scientist, tell others how to do this? Is this "bad" knowledge? If you write a paper and explain how to do something that is then used for a criminal purpose, do you have any liability? Where do we draw the line between what the public "needs to know" and what might cause harm? And who draws the line?[1]

From this story, you can see all the aspects that can confound, promote, or destroy scientific research. Is there knowledge that we should not know? Should there be limits on what we can do? If we do not understand and work within these constructs, we may be dooming ourselves to failure.

---

[1] Michael J. Selgelid and Lorna Weir, "The mousepox experience: An interview with Ronald Jackson and Ian Ramshaw on dual-use research", *EMBO Reports* 11, no. 1 (2010): 18–24, doi: 10.1038/embor.2009.270.

Biology and Genetics: Genetic Engineering

"Any sufficiently advanced technology is indistinguishable from magic."

—Arthur C. Clarke

GENETIC ENGINEERING

Genetic engineering is a way to add, remove, or alter the genetic makeup of a species in order to eliminate disease, produce a new organism, or increase "good" characteristics. Currently, the technology is mostly used to alter the genetic makeup of plants or, in some cases, animals raised for food. However, the potential is there for its use to manipulate the health or appearance of human beings.

We can produce a crop that is impervious to drought or a specific insect infestation or lack of sunlight. The list goes on and on. We can produce chickens with more white meat and beef with more taste. We have been doing this for many years. The concept is really not that different from selective breeding. Farmers for years have sought out animals with specific characteristics to breed to produce hardier, larger, more profitable animals. Modern genetics began with the work of Gregor Mendel in the mid-nineteenth century. His work with peas set the stage for the science behind nascent genetics. In 1953, DNA was first modeled by Watson and Crick to establish the familiar double helix structure. After that, we had to figure out how to remove or trans-

Shutterstock/Reinhold Leitner

fer genetic material. By 1982, scientists had managed to genetically modify a mouse, and by 1983, they produced a jumbo mouse by injecting it with human growth hormones. At this point, you should probably be asking why one might want a jumbo mouse.

After that, work began in earnest to manipulate the genes of various plants. This had great commercial appeal because it was usually done to produce hardier or tastier plants. You can use this technique to grow larger animals; you can use it to protect a crop from drought, pests, or virtually anything. In some cases, the issue was the ability to produce foods for distant markets. Bananas, for example, require a

Shutterstock/JacobSt

tropical or near-tropical climate to thrive. Scientists had to figure out how to design a hardier plant that could be easily transported. We also had to figure out how to eliminate the seeds (which we have basically done—all that's left are those tiny black spots). Unfortunately, plants that have been engineered in this way do not contain seeds that will germinate. Bananas are, today, essentially sterile. To grow new bananas, you must cut and plant a root. In the 50s, the most popular banana in the world was the Gros Michel variety, which essentially was eradicated by a fungus. Scientists have worked for many years perfecting the best tasting and the most transportable banana.

Today, for a number of scientific reasons, we only have one banana (unlike apples and other fruits that have numerous varieties). The banana today, the Cavendish, can be found virtually anywhere in the world.

You can also use it to grow certain pharmaceuticals within plants (a cheaper alternative to most pharmacology).

Harvard developed the "oncomouse," which could develop cancer, thereby giving us a viable animal for research and treatment. We have managed to grow a human ear on a mouse, which has unlimited potential for providing "normal looking" appendages for people who have lost theirs.

One of the reasons we were so slow to eradicate diseases like Hansen's disease (also known as Leprosy) was that man was one of the few animals that got the disease. Once we discovered that armadillos could also get the disease, it was easier to work on prevention and cure.[1]

In 1993, "U.S. Food and Drug Administration (FDA) approved Bovine somatotropin (bST), a metabolic protein hormone used to increase milk production in dairy cows for commercial use. Scientists determined which gene in cattle controls or codes for the production of bST. They removed this gene from cattle and inserted it into a bacterium Escherichia coli. This bacterium produces large amounts of bST in controlled laboratory conditions. The bST produced by the bacteria is purified and then injected into cattle."[2]

Shutterstock/Bruce Rolff

---

[1] "A brief history of genetic modification," *GM Education*, accessed March 30, 2015, http://www.gmeducation.org/faqs/p149248-a-brief-history-of-genetic-modification.html.

[2] Ibid.

By 1995, we had tobacco that could produce hemoglobin. We had corn and potatoes that had been injected by a bacterium Bt (Bacillus thuringiensis) to create resistance to certain pests. Dolly the sheep, our first cloned animal, was born in 1996 (more about cloning later).

The possibilities are endless. However, the science behind the ideas has also come to clash with many of our traditional values. A company can genetically engineer and patent a specific kind of seed. Said company can also use what is called "Terminator Technology." With this, a farmer cannot save the seed for the next planting; he or she must buy the seed each year from the company holding the patent. There have also been numerous lawsuits concerning crops that were "accidentally" cross-pollinated. You can be liable for damages if your crop that has not been genetically modified takes on the characteristics of a nearby genetically altered crop.

In recent years, genetically modified crops have been shown to be very successful. However, some countries have decided to ban them because of the uncertainty around the crop itself. You are dealing with future generations, and it is difficult to predict what might happen when we modify genetic material in agriculture.

If you feed a cow Bovine somatotropin (BST) hormone, for example, there is some question about the effect of BST on a small child drinking the milk. Americans have become more interested in organic crops that do not involve hormones. People will point to the simple fact that humans are much larger and that the onset of puberty continues to be younger and younger. The opposing side to this argument has to do with modern diet and better health care as the causative factor for this phenomenon. There is also evidence that people have stopped getting taller. The average height has allegedly stabilized in the last 50 years or so. Obesity, on the other hand, is a different proposition.[3, 4]

The earlier onset of puberty is a far more serious problem because of the possible connections between early onset of puberty and breast and uterine cancers. There is no definitive link, and there are certainly many possible mitigating factors, but the fact remains that we do not understand why girls of 8 and earlier are developing pubic hair and breasts. While many are quick to blame growth hormones fed to cattle, there is also evidence to establish that there are environmental endocrine disruptors causing the changes.[5, 6]

Shutterstock/SwissMacky

---

3 Agata Blaszczak-Boxe, "Taller, Fatter, Older: How Humans Have Changed in 100 Years," *Live Science* (blog), July 21, 2014 (08:58 a.m.), accessed November 11, 2014, http://www.livescience.com/46894-how-humans-changed-in-100-years.html.

4 Cynthia L. Ogden, et al., "Mean Body Weight, Height, and Body Mass Index, United States 1960–2002," Advance Data From Vital and Health Statistics no. 347, Division of Health and Nutrition Examination Surveys, Centers for Disease Statistics, October 27, 2004, http://www.cdc.gov/nchs/data/ad/ad347.pdf.

5 Elizabeth Weil, "Puberty before Age 10: A New Normal," *The New York Times Magazine*, March 30, 2012, http://www.nytimes.com/2012/04/01/magazine/puberty-before-age-10-a-new-normal.html?pagewanted=all&_r=2&.

6 Samim Özen and Şükran Darcan, "Effects of Environmental Endocrine Disruptors on Pubertal Development," *J Clin Res Pediatr Endodocrinology* 3, no. 1, (2011): 1–6, accessed November 11, 2014. http://www.ncbi.nlm.nih.gov/pmc/articles/PMC3065309/.

Asthma Period Prevalence and Current Asthma Prevalence: United States, 1980-2010

Current asthma prevalence, 2001-2010

Asthma period prevalence, 1980-1996

Percent

1980 1982 1984 1986 1988 1990 1992 1994 1996 1998 2000 2002 2004 2006 2008 2010
Year

The percentage of the U.S. population with asthma increased from 3.1% in 1980 to 5.5% in 1996 and 7.3% in 2001 to 8.4% in 2010.

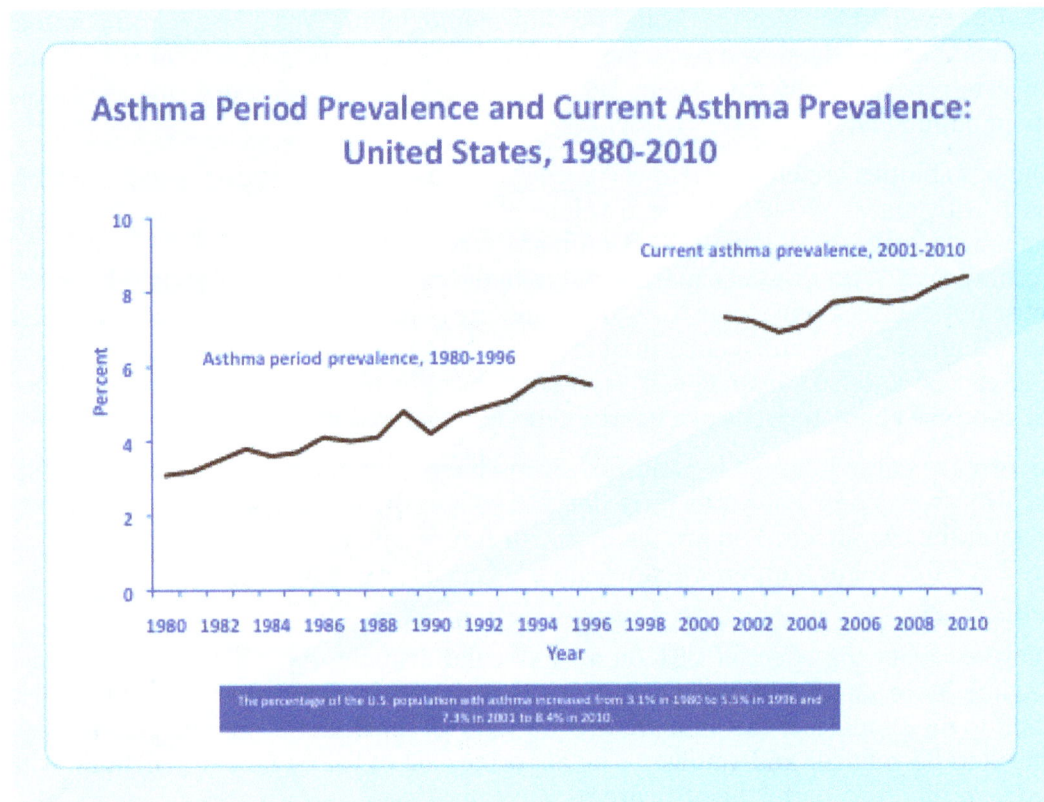

Centers for Disease Control and Prevention

Another problem has to do with the gigantic increase in asthma and allergy-related conditions. If, for example, a scientist decides that he wants a food crop to be hardier, he or she could conceivably choose something like a gene from a sunflower or some other hard-to-kill plant. The food labeling might not mention anything about it since it occurred on a genetic level. A child might be allergic to the added organism and be unaware of its existence in the food being eaten.

There is also the problem of routinely adding antibiotics to keep farm animals healthier. The current belief is that this fact, coupled with our overprescribing of unnecessary antibiotics for viral conditions, has led to antibiotic resistance. An antibiotic is only effective against a bacterial organism. It has no effect on a viral organism. We will discuss this later when we talk about medicine and health.

By 2000, scientists had managed to produce a potato that carries the vaccine against Hepatitis B. It can be imagined that, in the future, there will be no more vaccines; they will simply be included in your food source. However, the very fact that such a thing does not exist for human consumption over 10 years later is a powerful commentary on the problems with such development.[7]

Genetically engineering crops and animals may well hold the key to our future development as human beings. However, there are many possible negatives involved. If my company spends time and resources to develop a specific kind of fancy great tasting apple, do I have the right to make the seeds for that apple sterile? How else would I protect my costs and research? What could be the effect on

_____

[7] Julia Karow, "Potato Vaccine against Hepatitis B," *Scientific American*, last modified October 30, 2000, http://www.scientificamerican.com/article/potato-vaccine-against-he/.

third-world countries where saving seed may well be the difference between living and dying?

Today, genetic engineering is beginning to show us that we can possibly change the genetic makeup of a human being and that, perhaps, it will be the answer to future medical and health issues as well. It is within the realm of possibility that you could "engineer" your future child to have blue eyes or blonde hair or whatever else you might wish.

The idea behind genetic engineering (in human terms) is simple. The scientist needs to, first, map the genetic material and, second, add or remove what you wish. Theoretically, you could, in the future, simply remove some pesky diabetes gene and prevent your offspring from suffering from that disease. Of course, we know that it is not that simple (and probably will never be). However, the premise is amazing. Few parents would not jump at the chance to eliminate the possibility of severe or chronic disease in their offspring. There are, of course, people who are opposed for religious reasons. As with Frankenstein, it could be defined as "playing God." Both sides present difficult arguments. Unfortunately, on the science side, there is currently the possibility of causing even more havoc in future health. Removing a gene (or changing one or adding one) is a risky concept. We simply do not know enough about genetic engineering to provide safe, consistent results.

As with most technology, there is also the real probability of misuse. If you spent your life concerned with some specific characteristic of your own that you did not wish to pass down to your children, wouldn't you sooner or later try to alter that as well? And, of course, there is also the issue that this kind of "treatment" will probably be excessively expensive. If only rich people can afford to engineer characteristic "Q"—how is this different from Hitler's regime of Aryan blue-eyed blondes?

There is another aspect of the issue that surrounds genetic engineering that involves making an animal more "human," genetically speaking. While it sounds like an advertisement for a horror movie (and it may well be), the concept arose for several reasons. First, we might be able to better test pharmaceuticals on animals and remove the possibility of injury or death for human subjects. Of course, the animal rights activists do not agree whatsoever and, in fact, the legal issues only escalate from there. What kind of legal rights does something that is, perhaps, 40% human have?

We already have cows with enough human chromosomes to provide a possible source of antibodies to the Hantavirus. It is conceivable that we could create such an environment to prevent the spread of diseases like Ebola.[8]

In the second instance, we can conceivably "enhance" an animal body part (such as a heart or liver) to eventually use that body part for transplantation into humans. This concept, known as xenotransplantation, began with the idea of injecting substances into animal organs that would make them less likely to be rejected

Shutterstock/imredesiuk

---

[8] David Schultz, "Cows with Human Chromosomes Enlisted to Fight Hantavirus," *Science*, last modified November 26, 2014, http://news.sciencemag.org/biology/2014/11/ cows-human-chromosomes-enlisted-fight-hantavirus?ref=em.

when put into humans. Again, the notion, while meritorious on the face of it, has significant issues. Cross-species development is an area that could cause even more problems than it has solved so far.

The government is attempting to regulate any cross-species research, but there are no clear guidelines to date. Historically speaking, it has rarely ever been safe to assume that scientists will maintain ethical guidelines.

In later chapters, we will consider the aspect of cloning particularly as it relates to body parts. Genetic engineering and modification may be the direction of medical research and development for the future. Our current level of expertise, however, is not proven or safe enough for broad-scale applications.

DISCUSSION QUESTIONS

1. Do you think Dr. Frankenstein was wrong to try to do what he did? Why or why not?

2. What parallels can you see between the "science" of Dr. Frankenstein and research today?

3. What other examples can you think of where man has done the wrong thing for all the right reasons?

4. Do you think that the "monster" was inherently bad or that the perception of society made him "bad" (nurture vs. nature)

5. Given what we know today, if we could transplant some portion of brain tissue, do you think it would carry any kind of memories?

6. Do you think ALL scientific research information should be released? When should it not be? Why?

7. Do you think that we could create something that is "alive?" How would you define alive?

8. Are you worried about genetically modified foods? What are the pros and cons for you?

9. Do you think cross-species engineering is a good idea? Why or why not?

10. Should we abandon the concept of saving seeds for future generations in order to save money/time in food production? Why or why not?

LITERARY SOURCES—FICTION

Shelley, Mary, Frankenstein

RESOURCES AND REFERENCES

Learn Genetics: http://learn.genetics.utah.edu.
The Human Genome Project: www.genome.gov

Biology and Genetics: Cloning and Stem Cell Research

"Blind we are, if creation of this clone army we could not see."

—Yoda, Star Wars

CLONING

Cloning is based on the practice that you can make a genetically identical copy of a living organism. There are many reasons why this might be helpful, but there are equally as many reasons why it might be a dangerous idea.

There are three different types of artificial cloning:

1. Gene cloning
2. Reproductive cloning
3. Therapeutic cloning

Gene cloning creates a copy of a gene or segment of DNA. Basically, gene cloning involves taking the genes from one organism and implanting them into a vector. A vector can be bacteria, viruses, yeast, or another type of host organism. The vector allows the cell to multiply and make identical copies.

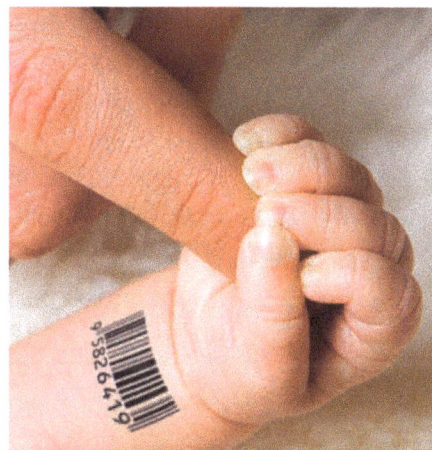

Thorsten Schmitt

Reproductive cloning is the idea of making an exact copy of an animal. To do this, researchers remove the genetic material from the egg cell of one animal. Then they remove a mature somatic cell from another and implant that in the "empty" cell. This allows the new cell to multiply and reproduce using the DNA from a different organism. As of 2015, scientists have managed to clone sheep, cats, deer, dogs, and many other animals. The first and most famous of these was Dolly the sheep, cloned in 1996. It is important to note that it worked after 276 attempts.[1]

There is also evidence that shows that the organs of these animals might be defective, which is not immediately apparent. Some die early (this may have been the case with Dolly).

Thus far, no humans have been cloned. Even though we may have the technology, there are severe ethical constraints on the idea of cloning humans. One of

---

[1] "Cloning Fact Sheet," National Genome Human Research Institute, last modified April 28, 2014, http://www.genome.gov/25020028.

Shutterstock/Pi-Lens

the big problems with this kind of research is the incidence of damage or disability. Many defective animals are produced before one arguably "perfect" animal is produced. This kind of cloning might be invaluable in the maintenance or reproduction of animals that are already extinct or could become extinct in the near future. Of course, there are interesting ethical repercussions from that idea as well. Do we really need saber-toothed tigers or wooly mammoths to return?

This type of cloning might also be involved in genetic engineering. Scientists may try to introduce "human" cells into animals in order to test pharmaceuticals, for example.

Therapeutic cloning is the idea that a scientist could, for example, create a body part when one is lost due to disease or injury. It is controversial because, at this time, it involves using embryonic stem cells. The source of this tissue would be from an egg cell that had only begun to divide and multiply. This kind of cell can potentially develop into any kind of cell. Using this technology could aid with the problem of organ rejection, for example. There is also the problem of organ transplantation. In 2015, over 100,000 people are on lists awaiting a transplant.[2]

Most "adult" stem cells represent only one kind of cell. Current research into this is beginning to show other methods of producing an embryonic-type cell. Without that, the embryo had to be destroyed in order to harvest the cells. New techniques are constantly being introduced to avoid the problem of destroying embryos.

If a scientist can create a specific type of cell in a test tube or growing medium, then they could use that artificial substance to test theories about disease and development. Scientists have already managed to grow human brain cells in a culture. Studying those cells in that medium could answer questions we have about ideas like mental illness or dementia.[3]

Shutterstock/Grzegorz Placzek

There is one other factor involved that needs to be considered. If we take, for example, a pair of identical twins, these twins would have the same genes. However, if they were raised in totally different environments, they may lose their obvious similarities. The reason for this has to do with epigenetics.

Epigenetics is the study of the factors that determine why some genes get expressed (or used) while others do not. Your genes may be hardwired into your body, but other factors can interfere with the expression of those genes. If one twin had a terrible diet, smoked,

[2] UNOS Fact Sheet, *United Network for Organ Sharing*, accessed March 28, 2015, http://www.unos.org.

[3] "Brain Organoids," MIT Technology Review, accessed March 28, 2015, http://www.technology review.com/featuredstory/535006/brain-organoids/.

drank to excess, and so on, he would most likely look different than his identical twin that did none of the above. One twin might end up taller, fatter, or have other characteristics that are not part of a specific genetic makeup. If external factors are radically different, the twins may no longer appear physically identical.

These external factors can also play an important role in the occurrence of diseases. Thus, we cannot simply study genes. We must also study environmental factors as well.

DISCUSSION QUESTIONS

1. Do you think we should clone humans? Why or why not?
2. Do you think we should use cloning techniques to "grow" a missing body part? Why or why not?
3. What is the difference between a clone and an identical twin?
4. Would a cloned human being be fully human?
5. How would different religions look at a cloned human?
6. Should we be cloning extinct animals? What is right or wrong about it in your opinion?
7. What would the religious approach be to such research?

RESOURCES AND REFERENCES

Cloning, http://www.genome.gov/25020028.

Stem Cell Basics, http://stemcells.nih.gov/info/basics/pages/basics1.aspx.

What is Cloning?, http://learn.genetics.utah.edu/content/cloning/whatiscloning/.

Human Cloning, http://www.ama-assn.org/ama/pub/physician-resources/ medical-science/genetics-molecular-medicine/related-policy-topics/stem-cell-research/human-cloning.page?

United Network for Organ Sharing, http://www.unos.org

Bringing the Back to Life, National Geographic, http://ngm.nationalgeographic. com/2013/04/125-species-revival/zimmer-text.

Cloning the Tasmanian Tiger, Part 1 https://www.youtube.com/ watch?v=sJL-_HutrsA

Part 2 https://www.youtube.com/watch?v=yoyzqcIrCV0

Human Clones: Through the Wormhole https://www.youtube.com/ watch?v=dzJx2dl8MA8

The Story of Dolly the Cloned Sheep | Retro Report | The New York Times https:// www.youtube.com/watch?v=tELZEPcgKkE

Epigenetics https://www.youtube.com/watch?v=kp1bZEUgqVI

Biology and Genetics: Biotechnology

"We are stuck with technology when what we really want is just stuff that works."

—Douglas Adams, The Salmon of Doubt

BIOTECHNOLOGY

Biotechnology is the science of connecting biology and living processes to real products with practical uses and applications for humans. Often when a scientist discovers something new, that knowledge may remain in the simple pure research mode. In some cases, however, we can translate that knowledge into applications that will help us or change the way we do business in the real world.

Biotechnology takes scientific research and applies it to the real world. While economics might play a part in whether a scientist gets funding for his or her work, in the biotechnology industry, the practical application of research and the cost-effectiveness drive the development of a final product.

In a sense, our first incidence of biotechnology in the real world might be something so simple as finding animals and putting a cage or pen around them so that they will be available for human use. When we first began to control and use methods to improve a crop to make it better (or easier) for human consumption, that also could be considered biotechnology. Someone, somewhere, first figured out how to add yeast to something to produce a different, fermented product.

Sometimes, research provides that "eureka" moment. An example of this might be the study of how our cells provide energy to our bodies—a strictly biological research concept. A scientist in an isolated lab might have looked at this concept in a new and different way and decided that perhaps we could harness that process and use it to power some device. Or perhaps the first step was to define the problem: a battery weighs x amount and if you are going to use it inside the body or in something small, then you have to figure out how to produce energy in a safe or tiny area. With biotechnology, you have to develop the process and then figure out whether you can afford to produce it.

Shutterstock/somchaij

Current worries about the decrease in fossil fuels have helped to drive the concept of "green" building. We can design a building that uses less fossil fuel by proper site

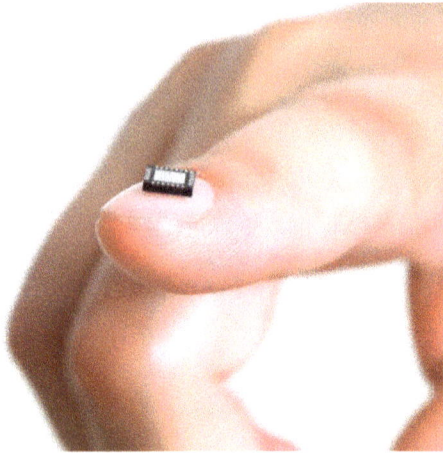

placement and the use of different materials. For that matter, biotechnology also drove the original ideas of how to build something that will last.

Using the concept of "biomimicry," a scientist might begin by studying how a limb actually functions both physically and bio-chemically. Biomimicry is the idea of using organic systems to produce mechanical systems. We might study a bird in flight to decide how to build an airplane wing, for example. The night vision of a bat might be used to figure out how to equip soldiers to see better in limited light conditions.

Aside from the aforementioned penning of wild animals, farmers also have learned how to add chemicals to soil to avoid the depletion of substances that might affect (or destroy) crops. With genetic engineering, we have seeds that are hardy enough to grow in drought conditions. We have crops that are pest resistant. All of these are part of the biotechnology industries related to agriculture. We can design a crop with a higher yield, which can feed more people for less money.

In medicine, scientists are using genetics to develop new methods of disease treatment. The pharmacology industry is perhaps one of the largest in the area of using biotechnology to provide viable and cost-effective results. With biomechanics, we can now build artificial limbs that are controlled by a person's brain waves. We are closer and closer to having an artificial limb that is "human" in both form and function.

With cloning, it is possible that we could simply "grow" someone a new arm in the future. At some point, this might be a cheaper and better method of replacing an injured limb or organ.

In the military, biotechnology is essential for both winning wars and protecting equipment and soldiers. Research developed for military purposes can provide aid and assistance to people with disabilities. A soldier is human, and thus, he or she can become tired, need a source of fuel, be weaker than an obstacle, or have visual or mental shortcomings. Seeing those issues, the military can design biotechnologies to enhance basic human strengths. Similarly, that same equipment might help a person with spinal injuries walk again.

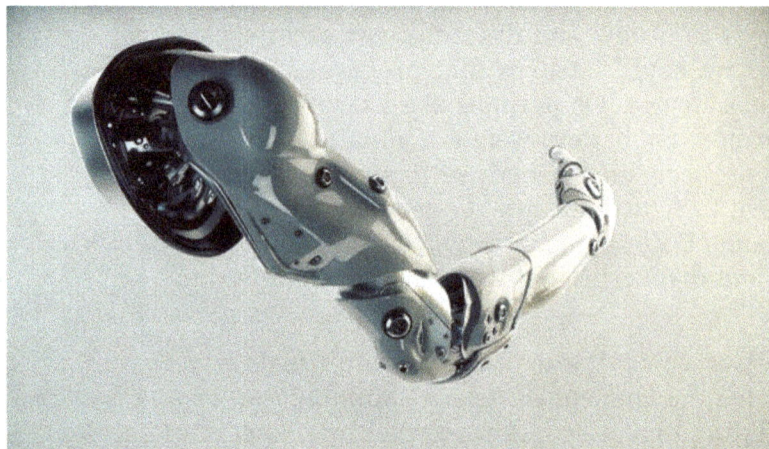

The Defense Advanced Research Projects Agency (DARPA) is the research and development arm of the U.S. Military.[1] The agency is divided into seven branches:

> Technology, Adaptability & Transfer (AEO)
> Biology, Technology & Complexity (BTO)
> Discovery, Modeling, Design & Assembly (DSO)
> Information, Innovation & Cyber (I2o)
> Decentralization, EM Spectrum, Globalization, & Information Microsystems (MTO)
> Networks, Cost Leverage & Adaptability (STO)
> Weapons, Platforms & Space (TTO)

Each of these branches is involved in cutting-edge research with military applications. Historically, the agency has pursued studies that many would consider science fiction. For example, DARPA has funded "mind control" research. DARPA's "robo-roach" could be considered a study in understanding brain injuries or how to use insects as spies.[2]

DARPA brought us the global positioning system (GPS) originally designed for military use. Other countries have been developing similar agencies.[3] The model for DARPA is very goal-oriented. Once an idea is generated, then DARPA will solicit and subsequently fund research on a given goal.

The military is not the only consumer of wearable technology. Products are constantly being designed that we can "wear" and accomplish certain tasks. Your cell phone, today, does many of the chores you once had to do with many different gadgets or applications. Disney, for example, is issuing a "special" wristband that allows the company to know about you before you appear in a theme park. A Disney "guest" walks in and a "cast member" greets you by name. One can be seated in a restaurant and they magically know where to bring your food.[4] All of this can be reassuring or terrifying. As with cellphone banking, just how much do they know about you? Mapping software on your phone is amazingly helpful in finding your way. It is also amazingly helpful for others to find you. That may be a good thing or a bad thing.

---

[1] Defense Advanced Research Projects Agency, accessed March 28, 2015, http://www.darpa.mil.

[2] Sharon Weinberger, "Ten Extraordinary Pentagon Mind Experiments," *BBC Future*, March 12, 2013, http://www.bbc.com/future/story/20130311-ten-military-mind-experiments.

[3] Conor Smith, "Canadian Innovation and the DARPA Model," April 10, 2014, http://natocouncil.ca/canadian-innovation-and-the-darpa-model/.

[4] Cliff Kuang, "Disney's $1 Billion Dollar Bet on a Magical Wristband," *WIRED*, March 10, 2015, http://www.wired.com/2015/03/disney-magicband/.

DISCUSSION QUESTIONS

1. What kind of ethical constraints do you think should be placed on a military agency like DARPA?

2. Do you think we should be pursuing "pure" research (research for the sake of enhancing what we know) or stick to research with practical applications. What are the pros/cons for both?

3. Give examples of historic scientific ideas that ended up being dangerous to the population at large.

4. Look over the DARPA website (www.darpa.mil). Find examples of current research and discuss whether you think it is a good idea and/or it would have practical application.

5. What kind of checks and balances do you think military research agencies like DARPA need to have?

REFERENCES AND RESOURCES

What is Biotechnology? http://www.ncbiotech.org/biotech-basics/what-is-biotechnology.

What is Biotechnology? https://www.bio.org/articles/what-biotechnology.

What is Biotechnology? http://www.whatisbiotechnology.org.

Learn About Greener Living, http://www2.epa.gov/learn-issues/learn-about-greener-living.

Upbin, Bruce, First Look At A DARPA-Funded Exoskeleton For Super Soldiers, Forbes, 10/29/2014, http://www.forbes.com/sites/bruceupbin/2014/10/29/first-look-at-a-darpa-funded-exoskeleton-for-super-soldiers/.

Regalado, Antonio, Military Funds Brain-Computer Interfaces to Control Feelings, A $70 million program will try to develop brain implants able to regulate emotions in the mentally ill, MIT Technology Review, 5/29/2014, http://www.technologyreview.com/news/527561/military-funds-brain-computer-interfaces-to-control-feelings/.

Weinberger, Sharon, Ten extraordinary Pentagon mind experiments, BBC Future, 8/12/13, downloaded 3/11/15, http://www.bbc.com/future/story/20130311-ten-military-mind-experiments.

Moon, Mariella, DARPA-funded mind-controlled robotic arm now works a lot better, Engadget, 12/17/2014, http://www.engadget.com/2014/12/17/darpa-mind-control-robot-arm/.

Smith, Conor, Canadian Innovation and the DARPA Model, The Atlantic Council of Canada, 4/10/2014, http://natocouncil.ca/canadian-innovation-and-the-darpa-model/.

Marcum, Maggie, Assessing High-Risk, High-Benefit Research Organizations: The "DARPA Effect" Study of Innovation and Technology in China, 1/2/14, http://igcc.ucsd.edu/assets/001/505314.pdf

Kuang, Cliff, Disney's $1 billion bet on a magical wristband, Wired, 3/10/15, http://www.wired.com/2015/03/disney-magicband/?ncid=newsltushpmg00000003

Sullivan, Mark, A brief history of GPS, TechHive, 8/9/2012, http://www.techhive.com/article/2000276/a-brief-history-of-gps.html.

Human Exoskeletons—for war and healing https://www.ted.com/talks/ eythor_bender_demos_human_exoskeletons

Bionic Legs Help Spinal Cord Patient Walk http://www.livescience.com/49379- bionic-legs-help-spinal-cord-patient-walk-video.html

New 2014 DARPA Building Real Life Terminators Military Robots Documentary & Discovery HD https://www.youtube.com/watch?v=9qCbCpMYAe4

Biology and Genetics: Nanotechnology

Grant: Wait a minute! They can't shrink me.
General Carter: Our miniaturizer can shrink anything.

—Fantastic Voyage

NANOTECHNOLOGY

In the movie, Fantastic Voyage, a famous person is nearly killed by an assassin. In order to save him, the doctors decide to put some scientists inside a submarine and then shrink it all (with the people inside) to a tiny particle that can then be injected into the body. They immediately, of course, run into problems with rogue blood cells, strongly beating hearts, sphincter valves, and all sorts of medical nightmares (or nightmares when you are tiny). The movie is almost ludicrous in its special effects. No one watching it realistically ever imagined that such a thing could actually happen.

Have we shrunk some people down to inject them into bodies? No. Do we have tiny submarines that can float about the body and find or fix a problem? Yes. Welcome to the world of nanotechnology.

Shutterstock/3Dalia

Nanotechnology is the idea of using extremely small things in scientific fields. Size is measured in the nanometer, which is one billionth the size of a meter.

The use of nanotechnology allows us to add tiny particles to just about anything. We can make baseball bats less likely to break, make bicycle helmets that are safer, make clothes that won't wrinkle. We can add nanoparticles that can make something water-repellent, smudge-proof, scratch-resistant. Nanoparticles are used in sunscreen and tooth fillings.

"How small is one nanometer? The typical width of a human hair is 50 micrometers. One nanometer is 50,000th of a hair width."[1]

When we first began our conjectures about the nanotechnology, we imagined that we could eventually build tiny submarines just like the one in Fantastic Voyage—molecular-sized machines that could perform a specific function much faster and more efficiently than its larger siblings. From there, we began to imagine creating drugs and drug delivery systems. The idea of combining nanoparticles with other

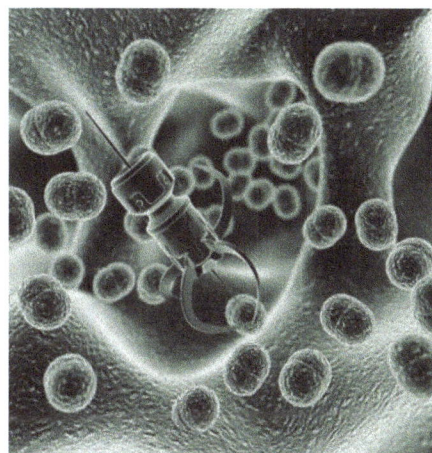

---

[1] Zhong L. Wang, "What is Nanotechnology?", *Professor Zhong L. Wang's Nano Research Group*, accessed March 28, 2015, http://www.nanoscience.gatech.edu/zlwang/research/nano.html.

"Although modern nanoscience and nanotechnology are quite new, nanoscale materials were used for centuries. Alternate-sized gold and silver particles created colors in the stained glass windows of medieval churches hundreds of years ago. The artists back then just didn't know that the process they used to create these beautiful works of art actually led to changes in the composition of the materials they were working with."[2]

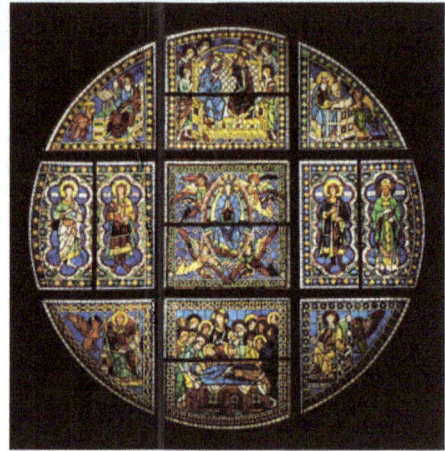

Shutterstock/Dmitriy Yakovlev

substances was a later development. To truly "be" nanotechnology, it must involve very small particles that can be controlled.

Current Uses of Nanotechnology Include:[3]

Boat hulls

Sporting equipment

Cosmetics

Sunscreen

Filling holes in bones

Dental implants

Anti-scratch coatings for glasses

Additives to medications to make them more absorbable/quicker acting

Computer chips

Stain- or rain-resistant clothing

Insulators to prevent energy loss

Air purifiers and filters

Specialized paints

New TVs, cameras, and cell phones use organic light-emitting diodes (OLEDs)

New battery technology

Fuel and solar cell

Stronger, less breakable car bumpers

Catalytic converters

Keeping our food safe (containers with nanoparticles)

---

[2] "National Nanotechnology Initiative," *United States National Nanotechnology Initiative*, accessed March 28, 2015, http://www.nano.gov.

[3] "Frequently Asked Questions," *United States National Nanotechnology Initiative*, accessed March 28, 2015, http://www.nano.gov/nanotech-101/nanotechnology-facts.

Every single day, the average person comes in contact with some variety of nano-technology. The future of this science is even more promising, particularly in medical applications. The "nanosubmarine" is already being tested. It can be used to find a tumor and to deliver medication directly to a specified area, which could conceivably reduce or eliminate our current methods of cancer treatment involving a broad spectrum approach that can damage surrounding tissue. Nanotechnology could also discover minute changes that could develop into cancer and treat them accordingly. At this point, it is basically still in the development/experimental phase, but the possibilities are real and likely to emerge within the next decade.[4]

Research by Dr. A. T. Charlie Johnson, a physics professor at the University of Pennsylvania, has produced a nanosensor that can "smell" cancer cells on the skin before melanoma itself is visible.[5] A person cannot smell at the level required, but a nanoparticle can.

All of these applications are growing exponentially on a daily basis. A nanobot could be "preloaded" with insulin and administer it as needed for a diabetic patient. Security cameras could be smaller than a pinhole and aid in crime detection. The science of nanotechnology evolves every day.

---

[4] Alliance for Nanotechnology in Cancer, "Where It Stands Now," *National Cancer Institute*, accessed March 28, 2013, http://nano.cancer.gov/learn/now/.

[5] Monica Rozenfeld, "Sniffing for Cancer," *Institute of Electrical and Electronics Engineers*, December 20, 2013, http://theinstitute.ieee.org/technology-focus/technology-topic/sniffing-for-cancer.

DISCUSSION QUESTIONS

1. Would you let a doctor inject a nanobot submarine in you? Why or why not?

2. What kind of restrictions should be made, in your opinion, on this kind of technology?

3. Can you see a scenario where this kind of technology could be easily weaponized? Give some specific ideas.

4. If we've had this technology, why hasn't it been used more to prevent cancer research? Can you find examples of problems where it has been attempted?

RESOURCES AND REFERENCES

Wang, Zhong L., Professor Zhone L. Wang's Nano Research Group, http://www.nanoscience.gatech.edu/zlwang/research/nano.html. www.nano.gov.

Nanotechnology Timeline, http://www.nano.gov/timeline.

Center for Responsible Nanotechnology, http://www.crnano.org/whatis.htm.

Nanotechnology Products, Applications & Instruments (List), http://www.nanowerk.com/nanotechnology/nanomaterial/products_a.php.

National Nanotechnology Infrastructure Network (NNIN), http://www.nnin.org.

Nanotechnology Applications, United States Department of Labor, Occupational Safety & Health Administration, https://www.osha.gov/dsg/nanotechnology/nanotech_applications.html.

Consumer Products Inventory, The Project on Emerging Nanotechnologies, http://www.nanotechproject.org/cpi/.

NCI Alliance for Nanotechnology in Cancer, National Cancer Institute, http://nano.cancer.gov.

Estimated Timeline for Select Nanotechnology Cancer Applications, http://www.nanotechproject.org/process/assets/files/2821/cancer_apps_timeline.pdf.

Rozenfeld, Monica, Sniffing for Cancer, IEEE, 12/20/2013, http://theinstitute.ieee.org/technology-focus/technology-topic/sniffing-for-cancer.

What is Nanotechnology? https://www.youtube.com/watch?v=TuljCWV6gLU

What is Nanotechnology? Introduction to Nanotechnology: https://www.youtube.com/watch?v=xlYIex2TF5g

Biology and Genetics: The Human Genome Project

"What more powerful form of study of mankind could there be than to read our own instruction book?"

—Francis S. Collins

(Source: From White House Announcement of the Completion of the First Survey of the Entire Human Genome Project, broadcast on the day of the publication of the first draft of the human genome. Quoted in transcript on the National Archives, Clinton White House website, "Text of Remarks on the Completion of the First Survey of the Entire Human Genome Project" (26 Jun 2000).)

THE HUMAN GENOME PROJECT

The Human Genome Project was a multinational collaborative to map out and understand all of the genes of every human being. The word "genome" is used to describe all of our genes collectively.

The Human Genome Project was a concerted effort built upon research done for many years and officially begun in 1990. It was projected to be completed in 15 years. Most of the first 5 years were spent developing techniques to achieve the goal. However, by 2003, we had completed sequencing the human genome.[1]

Today, if you visit the Mary Brown Bullock Science Center on the campus of Agnes Scott College in Atlanta Georgia, you will see a three-story mural depicting the DNA of Agnes Irvine Scott (for whom the college is named). Her great-great-great granddaughter provided the DNA sample. You can send off a kit and discover your own DNA. Such kits have many diverse uses including genealogical research and health.

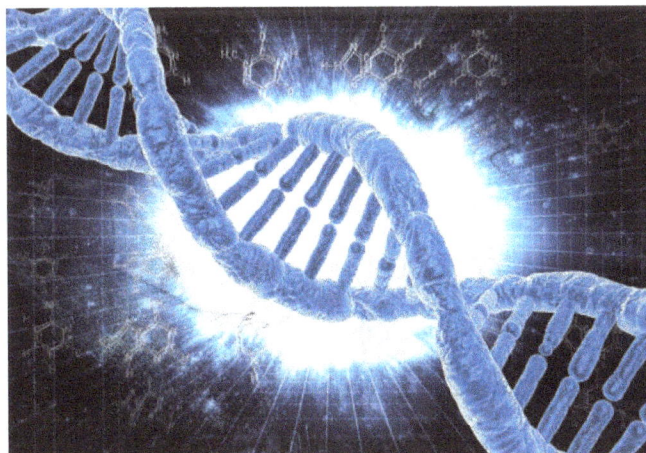

Shutterstock/vitstudio

---

[1] National Genome Research Institute, *"The Human Genome Project Completion: Frequently Asked Questions,"* last modified October 30, 2010, http://www.genome.gov/11006943.

Shutterstock/isak55

Why does it matter? What can we do with this research?

"The human genome contains approximately 3 billion of these base pairs, which reside in the 23 pairs of chromosomes within the nucleus of all our cells. Each chromosome contains hundreds to thousands of genes, which carry the instructions for making proteins. Each of the estimated 30,000 genes in the human genome makes an average of three proteins."[2]

Scientists have long realized that many diseases and abnormal conditions are genetic. Diseases like Huntington's disease, cystic fibrosis, hemophilia, and sickle cell anemia, to name but a few, clearly have a genetic component. Down syndrome involves a change in genetic material. It became apparent that if we could map the human genome completely, then we could possibly find the area that was "different" or abnormal. If we could find the aberration, then we could possibly manipulate that gene and change the health outcome. Some cancers also have a genetic aspect. Certain kinds of breast cancer have been tied to the BRCA1 gene first identified in 1990.[3]

Mapping the human genome became a major step in the fight to prevent the occurrence of certain kinds of disease. The concept of "mapping" means that scientists had to create a map of the exact sequencing or order of all the genes in human beings. We tend to think of the 23 pairs of chromosomes—a number and a fact most people learned in elementary school. When we talk about building a map, however, this involves the task organizing 30,000 genes and putting them in the correct order. From a genealogical standpoint, you can measure when characteristics change. From a medical standpoint, you can now find the point on the map where something might have gone wrong.

In the given illustration, you see a series of letters. Each letter stands for a base:

A: Adenine

T: Thymine

G: Guanine

C: Cytosine

We now know that humans have a specific sequence of these letters, following a specific layout. We also know that other organisms have the same letters, but different numbers of them and different layouts. We have, so far, only figured out a small portion of what these pairs might actually mean. We know, for example, that some of the coding here reflects things (like having a tail) that are no longer important to us. There are also indications that many generations ago, certain viruses might have invaded our species and changed this coding. One other interesting fact is that having more genes does not make us "smarter" or indicate that we are a higher order of being. Some simple organisms have more than we do.

---

[2] Ibid.

[3] "Genome: Unlocking Life's Code," *Timeline of the Human Genome*, Smithsonian National Museum of Natural History, accessed March 28, 2015, http://unlockinglifescode.org/timeline.

We can also sequence the gene of bacterium that causes diseases. This gives us a new way to possibly change the underlying structure of an organism to make it less lethal. It can also help in the development of vaccines to prevent the disease from occurring.

The Human Genome Project is one of the few such projects that provided results that were available to everyone without regard to copyright, national origin, or any other aspects of "ownership." More than 20 countries were involved in the research. It is easily one of the most important discoveries of mankind—the first attempt at figuring out our own "recipe."

There are obviously many kinds of ethical implications from this research. The original project was very careful to factor those concerns into the study. Consider, for example, what it would mean if we—as a human species—could know in advance what terrible health outcomes might await us? If you knew that you carried a gene for X type of cancer, how would that change your life? At this point, while it is valid to establish the map, many people might opt not to actually "know" if there is nothing that could be done about it. Some people have even chosen to never have children knowing the possible outcomes from a genetic disease.

In addition, the idea of genetic manipulation is only in its infancy. If you get ill today and you take medicine to "cure" that, as is the case with many diseases and that is the end of the problem. However, if a scientist can manage to alter or change or remove a gene that causes a problem, we have little information now about what that change might look like in years to come for future generations. We are only beginning to learn what some of these genes do. We still are pretty clueless about a large number of them.

DISCUSSION QUESTIONS

1. If you could change your own genetic structure to make you smarter, healthier, better, would you? Why or why not?

2. If you could change your own genetic structure to eliminate a disease in your future offspring, would you? Why or why not?

3. Would you want to predictively change the appearance or characteristics of a future offspring (blue eyes, red hair, etc.)?

4. If you only had a limited amount of funding, where would you put it? Would you choose predictive DNA research or research into curing diseases or preventing them by more current methods (like vaccination)?

5. What kind of situations could occur with DNA research that might have bad or disastrous results?

RESOURCES AND REFERENCES

www.unlockinglifescode.org (free images).

Timeline of the Human Genome,

http://unlockinglifescode.org/timeline.

www.genome.gov.

http://web.ornl.gov/sci/techresources/Human_Genome/index.shtml.

Understanding the Human Genome Project—A Fact Sheet, http://www.nlm.nih.gov/medlineplus/magazine/issues/summer13/articles/summer13pg15.html.

A Decade Of The Human Genome (BBC Documentary) https://www.youtube.com/watch?v=Fgq-XoyorWY

How to sequence the human genome – Mark J. Kiel https://www.youtube.com/watch?v=MvuYATh7Y74

3 Sad Surprises: The Human Genome https://www.youtube.com/watch?v=F5LzKupeHtw

Cracking the Code Of Life | PBS Nova | 2001 https://www.youtube.com/watch?v=_IgSDVD4QEc

Human Genome Project Animation https://www.youtube.com/watch?v=UfsHl5_2rMw

Education: Introduction

"All of a sudden, we've lost a lot of control," he said. 'We can't turn off our Internet, we can't turn off our smartphones; we can't turn off our computers. You used to ask a smart person a question. Now, who do you ask? It starts with g-o and it's not God. . . ."

— Steve Wozniak

Fifty years ago, many college-level libraries were "closed" systems. If you wanted a book or magazine, you first had to go to the card catalog, find the title of the book you wanted, laboriously write the info down, then hand it to an officious person behind a desk. That person would then go get you the book (or, far more commonly, tell you that it was checked out). The burden of this kind of learning is that you had to have a decent sense of what you wanted or needed in order to find out anything. Even if you had an "open" library where you could search the shelves, you also needed to understand the organization of the information available in order to find materials.

Today, there is something of an educational backlash among students who no longer see the point of memorizing any factual information. With almost universal access to the Internet, most factual information can immediately be located. Some schools no longer teach cursive writing; why bother when no one uses it anymore? Math may very well go by the wayside next. If calculators are ever present, then why bother learning any math at all? We already have technology that can maintain and compute the information faster than most humans.

As we discussed earlier, the very abundance of information coupled with the constant multitasking may have an impact on developing brains. Either we will learn to effectively multitask or we will stop learning. If all information is immediately available, why do we still need to "learn" anything other than physical tasks? Of course, it could also be argued that we might not need to learn those either; such tasks might be more easily and quickly done by sentient robotics.

Unfortunately, the amazing access to information known as the Internet has several dark sides. First off, there is the question of whether everything actually should be accessible. Do we want to encourage censorship? Do we want to prevent a young child from seeing something that might be damaging? Who will decide just what they can or cannot see?

A far more insidious problem is the simple abundance of information. The Internet has brought us "water-bug" learning. If you think you need to know something, you decide what to put in the search bar. You skim through your 5,000,000+ responses (usually selecting one that reinforces what you already believe), you skim that and voila! Instant learning.

We are drinking the Kool-Aid.

Education: The Internet

"Computers are useless. They can only give you answers."

—Pablo Picasso

The Internet is an amazing idea. From a small network of interlocking military computers, we have come today to an almost universal access to information. You no longer have to find something in your card catalog or move with trepidation and begging to the librarian's desk for the one book that may or may not help you. We have moved past the idea of getting information and now mostly question just how fast we can get it. The question of just how valid said information might be is only part of the problem. The Internet can, to a certain extent, be a controlled entity. If your government does not want you to see something, it is conceivable that they could prevent you from seeing it.

Shutterstock/Maxx-Studio

Problem 1: Search Algorithms

You still have to decide what to search for. If you ask for the wrong thing, you get a possibly wrong answer. If you need to know why a teenager commits suicide, you might start with suicide, then realize you might need to qualify that somehow, and so on. You might start with the wrong concept for searching and believe you have received an answer when better searching might produce better results. You also routinely receive millions of responses. Many people scan the first page and go from there, clicking wildly on links. As we saw in the earlier chapter on filter bubbles, this might be automatically tailoring your response to what you already know or believe. Remember, also, that many words have many different meanings.

Problem 2: Just Exactly How Much Do You Actually Read?

When queried, most people respond honestly that they actually don't read ALL of any articles online—or, if they do, it is a rare occasion. If you are doing research, you might find a few, copy the citations, and move on. Technology has a way of

A student once asked me what she needed to connect her laptop to the school network.

The part that she needed was called a "dongle." I told her that and suggested she look it up online. The next week, when she returned to class, she was so upset, she was shaking. The word "dongle" is not just used for that specific computer part. She couldn't figure out just what she needed so she switched to "image" view. After that, we had a new lesson on how to conduct safe searches. Do not, as they say, try this at home.

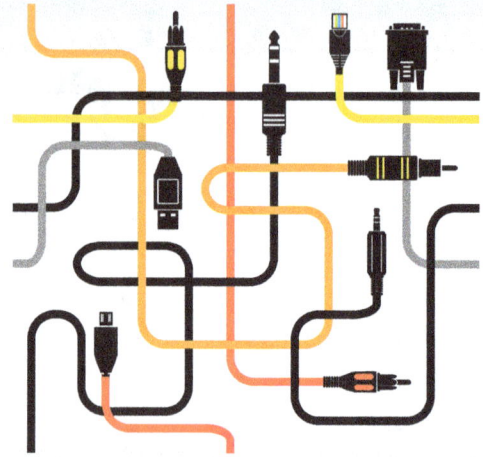

Shutterstock/tovovan

measuring how often you click on a certain page, whether you click on other pages on the site, and so on. Unfortunately, today, most people land on a page from a search engine (and thus do not always click around on a site). For example, we can measure that you "clicked through" from an ad on one page to a product on another. Counting clicks was our new improved method for measuring interest. One downside to this was, of course, that decent web pages are now covered with ads, flashing signs, videos, and other distracting messages. This is like clicking around on channels on your TV during a commercial, but on steroids. However, the most important issue is not our near universal access. The most important problem is whether we actually read ALL of any article. According to Chartbeat (a company that specializes in measuring website views), 55% of people spend fewer than 15 seconds on a page.[1] While you can actually measure what topics interest people, most Internet measurement is based on whether you access something at all. We also tend to think that people actually read something in its entirety before we share it (also not true).[2] You can actually measure whether someone scrolls down an article (implying that they might actually be reading all or any of it).

Chartbeat also can measure just how many just click through and move to another source, how many scroll, how many tweet or share before actually reading.[3] All of this implies that perhaps our "attention" is getting a bit unfocused, which brings us back to our earlier articles about just how we learn at all.

Problem 3: Fact or Fancy?

In a recent research paper by Google, it appears that Google has actually developed a type of algorithm that can rate information on the Internet. Simply put, it

---

[1] Tony Haile, "What You Think You Know About the Web is Wrong," *Time Magazine*, last modified March 9, 2014, http://time.com/12933/what-you-think-you-know-about-the-web-is-wrong/.

[2] Ibid.

[3] Farhad Manjoo, "You Won't Finish The Article: Why People Online Don't Read to End," *Slate*, last modified June 6, 2013, http://www.slate.com/articles/technology/technology/2013/06/how_people_read_online_why_you_won_t_finish_this_article.html.

can tell you whether something is true or not. Most people seem to think this is an amazing concept that should be implemented. At this point, it is only in the research and development phase.[4] The idea is based on an older product of Google known as "Knowledge Vault" where Google stored "a billion facts so far."[5] The more recent research takes that information and measures the accuracy of various websites. Some would welcome such information. However, the problem is not one of the assimilation of basic or "true" facts.

Slavoljub Pantelic

The problem, when it comes to the Internet is that so much of what we believe is based on faith, not facts. We will cover the various kinds of bias later, but most people will accept as "fact" something that they already believe regardless of whether it is true or not. Imagine that you have a sore throat. Your mother, your grandmother, all your relatives tell you to do X. You look it up and a known person of unknown credentials says you must go to the doctor. Add your faith in your family, your prior experience with doing what someone you trust tells you, and the fact that you don't have health insurance and what are you going to do?

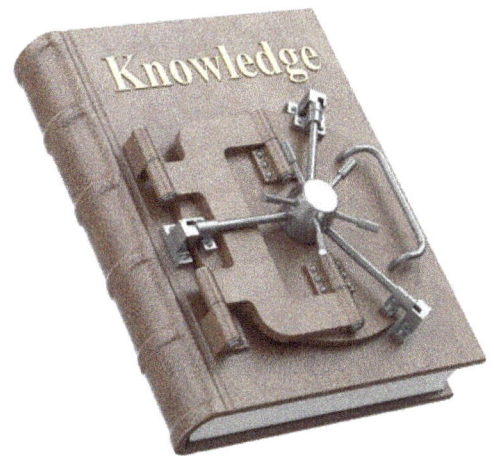

Problem 4: Are we better informed?

In a Pew Research survey from 2014, people overwhelmingly believe that the Internet has made them better informed.[6] On the face of it, this would seem valid since we can now instantly access a great deal more information than we once could—and we can do it immediately. The problem, however, is in how we use or assimilate the information that we find. The combination of our own bias and our own sense of "I don't need to know it, I just need to find it" causes us to lose our own ability to synthesize what we see, hear, and read.

Problem 5: How do we learn and do we need to?

If you remember from elementary school, you could have had a test with a question like "When was the War of 1812?" I'm sure you really liked those questions. Aside from the fact that the answer was in the question, it also involved the simple memorization of a fact (which, of course, was easier for some people than others). You probably (like most people) hated discussion or essay questions that

---

4 Caitlin Dewey, "Google Has Developed a Technology to Tell Whether 'Facts' on the Internet are True," *The Washington Post*, last modified March 2, 2015, http://www.washington post.com/news/the-intersect/wp/2015/03/02/google-has-developed-a-technology-to-tell-whether-facts-on-the-internet-are-true/.

5 Chris Mooney, "The Huge Implications of Google's Idea to Rank Sites Based on Their Accuracy," *The Washington Post*, last modified March 11, 2015, http://www.washingtonpost.com/news/energy-environment/wp/2015/03/11/the-huge-implications-of-googles-idea-to-rank-sites-based-on-their-accuracy/.

6 Kristen Purcell and Lee Rainie, "Americans Feel Better Informed Thanks to the Internet," *Pew Research Center*, last modified December 2014, http://www.pewinternet.org/2014/12/08/better-informed/.

Shutterstock/scyther5

required a bit more from you. Today, clearly, most people do not need to have memorized the dates of a given war when such information is readily available. You can change the question to "Why did the War of 1812 occur?" That is a bit more problematic, although you can certainly search for "causes of War of 1812" if you wish.

The Internet is changing the very way we learn. The old model of laying out facts and then "connecting the dots" has been taken over by looking up the facts and the dots and then believing that you know something. In a paper from 1999, two Cornell researchers, David Dunning and Justin Kruger, did a study on how people measured up in humor, logic, and grammar. What they discovered was a depressing paradox: those who are measured to be incompetent in humor, logic, and grammar tend to reach mistaken conclusions AND they don't know it. Known now as the Dunning–Kruger effect, this has serious implications when connected to learning. This could be a factor with our current issues in polarization. It doesn't matter if you attempt to "prove" something using facts; people have a basic "faith" in their own beliefs and there is little that can change that. The Internet has made this worse. If you only click on articles related to your own belief structure, then you are constantly reinforced in your very "rightness." It is almost easier to reject science or any other idea that doesn't fit in your worldview.

Problem 6: Collective learning

Once upon a time, you had an educational environment where you had a teacher and a class. The teacher, hopefully, provided decent factual information and helped you to figure out what to study and how to learn what you needed to know. Your "table" would sometimes have what came to be a "group" project. The idea behind this theory was to help you get along with others (part of the "good citizenship" agenda of public school). Today, the possibility exists of a group of children sitting and studying independently with electronic devices. They might be learning valid information but they are not learning how to get along with and/or manipulate their peers. One of the problems with the spread and encouragement of diversity has always been that limited exposure tends to equal poor knowledge.

Shutterstock/Andresr

While the Internet can bring you pictures of tribal life in some other country, it cannot provide you with the empathetic skills you might need to actually "know" another culture. While "group" learning has its pros and cons (no college class has ever been in favor of having a group project), it does equip you with a means of real conversation and the possibility of learning from the experience of another.

On the face of it, it would seem that the Internet could do that far better. In one sense (the very multitude of available information) it can; on the other hand, it cannot impart to you the feeling of holding hands with another. The Internet, in that sense, is behaving autistically, for want of a better expression. It is not emotionally involved with you and thus can lose the depth of human connection you might otherwise be able to experience.

Shutterstock/Rawpixel

DISCUSSION QUESTIONS

1. Do you think that your own education prepared you to work and live in the "Internet" world? Why or why not?
2. Do you think small children should be using electronics in early grades?
3. Do you read all of an article you click on? Does it matter?
4. Do you think video games have a place in education? How could they be used and/or where could they be problematic?
5. How does your own "learning" style fit with the Internet?
6. Do you think that cooperation is a valid part of your education? Why or why not?

RESOURCES AND REFERENCES

Dong, Xin Luna, et al., Knowledge-Based Trust: Estimating the Trustworthiness of Web Sources, http://arxiv.org/pdf/1502.03519v1.pdf.

Jeffries, Adrianne, You're not going to read this but you'll probably share it anyway. The Verge, 2/14/15, http://www.theverge.com/2014/2/14/5411934/youre-not-going-to-read-this.

Kruger, Justin and Dunning, David, Unskilled and unaware of it: How difficulties in recognizing one's own incompetence lead to inflated self-assessments. *Journal of Personality and Social Psychology*, 77(6), Dec 1999, 1121–1134.

Education: Critical Thinking

"Logic is the beginning of wisdom, not the end."

—Leonard Nimoy (Spock)

Most human beings are a curious mixture of rabid belief and healthy skepticism. With the abundance of information available today, exactly how is someone to figure out just what is true? A hundred years ago, you could look into a printed document and assign it a certain veracity simply because it WAS a printed document. There was power in something that someone went to the trouble of writing and promoting. While poor information certainly existed then, the very fact of a printed book tended to give it some degree of import.

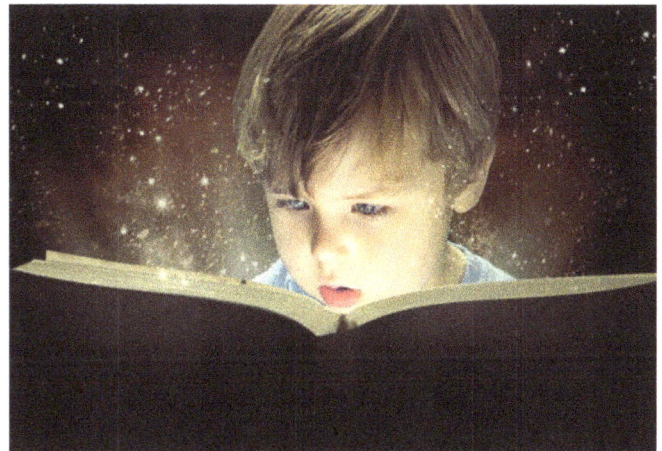

Shutterstock/conrado

Today, that is no longer true. Anyone can self-publish pretty well anything. In addition, of course, the Internet has brought us so many choices that it is very difficult at times to figure out just what is the truth (if that is even possible). In some ways, that is wonderful, but in some ways, it causes even more polarizing behavior and actions. Most people, when asked, truly believe that they can read information online and mentally verify its factual basis. This differs age to age. Some people, for example, may believe that a website has "power" or validity simply because the website itself looks professional. We tend to denigrate websites with lots of flashing ads or websites that look homemade. This is a way of attaching value to the cost of the website ("if they had something important to say, they would throw money at it.").

The easiest way to dispel this idea is to look at websites produced by big pharmaceutical companies. They definitely "throw money" at them; they tend to be highly professional in appearance and are designed, in a sense, to promote a feeling of well-being and good health. "You have X condition, here are all the

Shutterstock/PureSolution

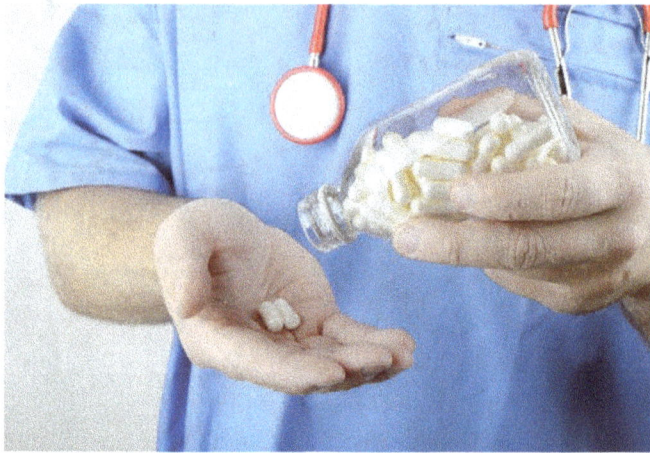

Shutterstock/Rob Hyrons

symptoms. Here's what the doctor says you should do." And, of course, the "doctor" is a "certified" professional and he or she thinks you should take X pill. Are they promoting "wrong" information? Not necessarily. They are, however, promoting their own agenda.

A savvy consumer is someone who reads reviews online, listens to the input of friends and family, and makes an informed decision about what to buy. The same can be said of the savvy knowledge consumer online. There is a set of questions to ask. While many people won't take the time to follow through and look at all of the possibilities, it is important to learn how to do so. No one wants to share that amazing video of a shark trying to bite a navy guy escaping into a helicopter only to find out that it was faked. Just as no one will ever get "that" email saying that you have been left x million dollars by an African prince.

Abiola and Associates
Solicitors at Large
Abuja

Dear Mr. Johnson,

I am Mr. Abel Abiola, a lawyer in my home country of Nigeria. I am the attorney of record for the late Chief Dr. Adeniji, who lived his life caring for others in our home country of Nigeria.

Chief Dr. Adeniji and his wife and only child were murdered by dissidents in the mountains of Chappai Waddi where they had gone to save the lives of many small children. Before his death, the kindly doctor requested that I hold his fortune in my keeping until such time as he could remove it from the country. He was very worried that the Nigerian military might choose to confiscate his holdings. His last will and testament states that I must secure someone outside the country to receive his fortune (currently valued at US$20 Million Dollars). He had only one brother who had moved to the United States many years ago and changed his name to Johnson. I have spent many years researching the brother and I believe he may be of kin to you. Since the bank has now given me 10 days to secure the next of kin. I believe that this money rightfully belongs to you.

Chief Dr. Adeniji wished that his money be spent taking care of orphans of war. If you are willing to take on this sacred trust, I believe that you could keep a share of 20% as your share of the fortune. As the administrator, I would retain 10% for my diligent work on this behalf.

I have all documents that we shall need to prepare this claim. If you agree, the money can be wired into the bank account of your choosing immediately once all papers are signed. This is a legitimate endeavor.

I know that you would wish to take on this sacred trust from your esteemed kinsmen.

Please respond to this email so that we can begin this transaction. I will need your telephone number so that we can discuss the details.

Best regards,
Mr. Abel Abiola

Most people (today) can recognize the falseness of this kind of email. Clearly, all do not, or perhaps the people perpetrating such a scam would quit doing it. In a case like this, you can recognize the warning signs (first of which is that you probably aren't going to get something for nothing).

But as we learn to avoid these types of scams, they tend to get more and more sophisticated. In order to protect ourselves, we need to develop good "crap" detectors before the need arises. To do that requires the application of logic. There are a number of ways people can attempt to convince you of the veracity of something they say. One of the primary problems here is that we often truly want something to be true and thus we ignore the evidence that it might not be.

"You know why I'm sure the sales chart will be great? Because this is the morale chart."

Shutterstock/Cartoonresource

"Logic! Good gracious! What rubbish!" (E. M. Forster)

A "fallacy," by definition, is a belief in something based on improper or unsound thinking. The kinds of fallacies are numerous, but here are the more common.

CORRELATION AND CAUSATION

Sometimes people look at something that happens and decide that X was the cause. Example: "Every child in the class today fell down at recess. Obviously this class is full of clumsy children." It ignores the concept that other factors (like a muddy playground) could have caused this instead of clumsy children. Any time you see a graph of unrelated items that appear to be in synch in the graph, this is usually the issue. One of the most famous is the "pirate" argument. As global warming increased, there were fewer pirates. You can plot this out on a graph with REAL NUMBERS. Therefore, we should all become pirates to prevent global warning.[1]

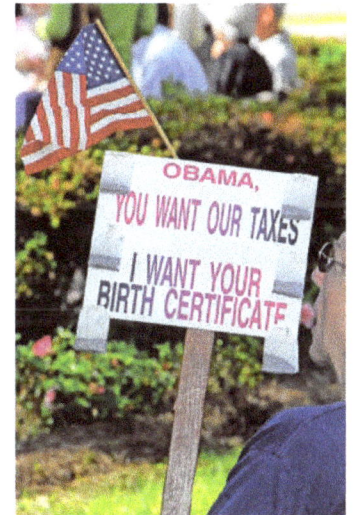

Shutterstock/Cheryl Casey

AD HOMINEM

This is fairly common, especially in politics. It is the idea of attacking the character of someone personally to avoid discussing the issue at hand. Example: "You should not agree with him about abortion because he is a criminal bad guy who does not believe in God." The idea is to destroy the credibility of the witness and have the person you are trying to convince be in solidarity with you.

---

1 Erika Andersen, "True Fact: The Lack of Pirates is Causing Global Warming," *Forbes Magazine*, last modified March 23, 2012, http://www.forbes.com/sites/erikaandersen/2012/03/23/true-fact-the-lack-of-pirates-is-causing-global-warming/.

Shutterstock/Rawpixel

AD POPULUM

With this kind of appeal, someone is basically trying to get you to buy in based on a patriotic-type appeal. Example: "If you were truly a good American, you would agree that we have to all drink soda." If you want to disagree, you would first have to wade through the "good American" part.

Shutterstock/Wuttichok Painichiwarapun

STRAW MAN ARGUMENT

In this kind of discussion, you hear something, and you respond with a tangential semi-related idea and use it to discredit the first argument. Example: Person A: "We need to cut the healthcare budget." Person B: "If we spent more money on healthcare, I would be healthier. Clearly we are less healthy because they cut the healthcare budget."

Shutterstock/Nevada31

RED HERRING

This is the idea of bringing up something else as a diversion so you quit thinking about the original issue. Example: "I think mercury in vaccines causes autism. Did you know that there is a ton of mercury in the fish we eat?"

Shutterstock/StockLite

ANECDOTAL

Sometimes people use their own family stories to prove a point (which often tends to involve correlation and causation). Example: "My grandfather smoked a pack of unfiltered Camels everyday and started his day with a whiskey toddy. He lived to be 98. Therefore, it is okay to smoke and drink." People hearing this might decide that smoking and drinking automatically make you live longer.

SLIPPERY SLOPE (THE DOMINO CONCEPT)

Often people will try to justify something and end up with something that is totally illogical and unrelated. Example: "If we let gay people marry, next people will want to marry monkeys." These are totally unrelated and the idea of marrying monkeys is both useless and ridiculous (not to mention offensive), since there is no connection.

Shutterstock/View Apart

GAMBLER'S FALLACY

A gambling addiction can begin with the simple concept that after you lose X number of times, the "odds" are that you will get a winning hand. Example: "I am taking a true/false test and I have 10 'false' answers. There's got to be a true answer soon."

Shutterstock/Fer Gregory

BANDWAGON

Useful especially in politics, this is the idea that you should think a certain way because "everyone is doing it." Example: "We need to support skinny-dipping because everyone is doing it."

Shutterstock/Fer Gregory

Shutterstock/PathDoc

AUTHORITY

Weak arguments can often begin with the idea that you should do something because someone famous, smart, important, and so on says it is a good idea. Example: "(Since I can't prove it any other way) Famous Dr. X told me that he believed that you did not ever have to diet to lose weight. Now, if you disagree, you are questioning an authority (not me)."

Shutterstock/Gwoeii

MOTHER NATURE KNOWS BEST

Usually used with product marketing, this is the idea that if we label something "natural," it is obviously better for us. Example: "Organic food is always best because it is more natural." On the face of it, it seems valid, but, in fact, "organic" food might not always be best (e.g., it tends to cost more).

Shutterstock/William Freeman

BEGGING THE QUESTION

Also called circular logic, this is the idea of using a statement to prove itself. Example: "We must believe the Bible because it says that we must."

BLACK AND WHITE

This concept is based on the idea that there are ONLY two choices. The implication is that if you don't do one, the other happens. Example: "We should all drive electric cars; if we don't the world will warm up and we will all die."

Shutterstock/Valentin Valkov

COMPROMISE IS BEST

We tend to believe that compromise is a good concept, and it usually is. However, when it comes to scientific "fact," it can only cause problems. Example: Person A: "I believe people cause global warming." Person B: "I believe the sun causes global warming." Person C: "Global warming is caused by people and the sun." The idea is not that person C would always be wrong; the idea is that sometimes there IS a specific cause and settling halfway between two ideologies isn't always helpful.

Shutterstock/Vuk Varuna

BURDEN OF PROOF

Sometimes people make a claim and then decide that the only way for you to disprove it is to "prove" you are wrong. Example: "I believe that the sun revolves around the earth. It is your job to prove me wrong." It is probably more helpful to prove your own version of things first before assigning the proof construct only to a negative.

Shutterstock/chrisdorney

CONFIRMATION BIAS

This is probably one of the most dangerous ideas when it comes to Internet use. This is the idea that you go into learning with a specific mindset and nothing you read can change it (and/or you only read things that agree with you). Example: "Since I believe in God, evolution could not be true and I won't believe anything you say."

Shutterstock/filmfoto

Shutterstock/Viacheslav Nikolaenko

MOVING THE GOALPOST (RAISING THE BAR)

This is the idea that "if only" the true test involved X condition, then I would have been proven right. Example: "Just because I can't guess what the next card will be in your test, it does not mean that I am or am not psychic. You need to have a quieter environment for a true test."

Shutterstock/DarkBird

NO TRUE SCOTSMAN WOULD EVER

Not as common, but still out there. This is the idea (similar to Ad Populum) that everyone who is truly "X" would never behave or act in a certain way. Example: "No true Scotsman would turn up his nose at haggis." Clearly, there are Scotsmen who don't eat haggis (hopefully).

Shutterstock/Ivonne Wierink

THE CHERRY PICKER

Sometimes a reporter will state a statistic and a politician will use it for a specific agenda. Example: "The government claims that 10% of children are now using marijuana while 20% are smoking cigarettes." The person with an agenda would then state: "We need to increase spending on the prevention of marijuana since such a high percentage of children use it." It is a way of ignoring data that doesn't support your agenda.

PUTTING IT ALL TOGETHER

Obviously, the first issue is that you actually want to know the truth of something. Unfortunately, many such concepts (shown earlier) are designed to keep you believing what you want to believe instead of what might be more correct.

KNOWING THE RIGHT QUESTIONS

1. Is it "unbelievable" or amazing?

2. Can you verify it on more than one well-known news program or reliable source?

3. Can you easily look up and verify someone quoted as "an authority?"

4. Is that person an authority in THAT FIELD?

5. If it is a "scientific" peer-reviewed study:

   a. Can you find the original research paper and verify that it is correct as reported?

   b. How many people were in the study shown?

   c. Was there a variety of age, sex, gender, and so on in the subjects of the study?

6. In a sequential argument, is every single step true?

7. Is it verifiable by another test or experiment? Do you accept/believe that someone else could ask the same questions and get relatively similar answers?

8. Does the article or source imply dire consequences if you do not accept or believe the premises stated?

9. Does the article attack someone who has a specific point of view? An ad hominem is to attack the person holding a point of view instead of the point of view.

10. Does the article confuse cause and effect? Example: "Yesterday, everyone in the playground fell at least once. Therefore, everyone on the playground was clumsy."

11. Does the article deliberately make fun of someone in order to strengthen the opposing point of view?

12. Does it back up something you already believe strongly? Do you really want it to be true?

    a. Spend some time verifying and studying the opposing points of view.

    b. Who sent the information to you? Do you believe they have an agenda?

    c. Consider the source (if a news item). Does that particular source always follow a specific agenda?

    d. If it is essentially reported by a political person or source, why are they reporting it now?

Hopefully, knowledge about HOW information is distorted will help someone to put together a set of information that is valid before deciding arbitrarily on a point of view.

DISCUSSION QUESTIONS

1. Do you honestly try to sort through information to find out what might be true?
2. For you personally, who exerts the most influence on what you believe?
3. Find examples of at least three of these concepts online. Provide links and explain why you think it fits.

REFERENCES AND RESOURCES

Hilarious graphs prove that correlation does not prove causation, http://www.fastcodesign.com/3030529/infographic-of-the-day/hilarious-graphs-prove-that-correlation-isnt-causation.

www.snopes.com.

The Nizkor Project, http://www.nizkor.org.

Google Correlate, http://www.google.com/trends/correlate. www.hoax-slayer.com.

Education: Electronic Media

"Lovers of print are simply confusing the plate for the food."

—Douglas Adams

BOOKS

The invention of the printing press is often marked as one of the singular technological moments in time. Before that, written knowledge had to be laboriously copied and was generally owned by religious groups or the wealthy. With the printing press, came literacy among the masses.

Many people continue to have an emotional attachment to a printed "thing" and cannot imagine shifting over to an electronic format. The spread of the Internet, however, was as important as the printing press. It set the framework for knowledge to be accessible by nearly anyone. There is a certain romantic appeal to a physical book, but many believe that it is totally outweighed by the very accessibility of an electronic book. You have, at your fingertips, (for a fee of course) access to billions of topics, books, materials, and other items of learning. There is no waiting and almost no filter. You want, you get it—even if it is two in the morning. With a small amount of light, you can read it in bed with an undisturbed sleeping spouse. You can highlight and find a passage far more easily. You can look up words, make notes, and search within an e-book. It may not make up for the emotional appeal of a real book, but there is some indication that electronic versions are the future. In 2010, the New York Times reported that Amazon had announced that e-books topped hardcovers.[2]

There is also an interesting corollary that e-book sales are also falling at various points. Since the audience would remain basically the same, there appears to be the possibility that many people are reading on devices that can do a

"He who first shortened the labor of copyists by device of movable types was disbanding hired armies, and cashiering most kings and senates, and creating a whole new democratic world: he had invented the art of printing."[1]

Shutterstock/rSnapshotPhotos

---

[1] Thomas Carlyle, *Sartor Resartus*, (*Oxford*: Oxford University Press, 2000).
[2] Claire Cain Miller, "E-books Top Hardcovers at Amazon," *The New York Times* (New York, NY), July 19, 2010, http://www.nytimes.com/2010/07/20/technology/20kindle.html?_r=1.

Shutterstock/vvoe

multitude of things (as in, why read when you can play solitaire?).[3]

There have been numerous studies trying to ascertain if reading an electronic screen will affect your behavior or eyesight. Any time new technology comes out, it is generally followed by studies that imply that doing X (reading or viewing any kind of electronic screen) will have dire consequences. We looked at some of those in the social media section, but it is interesting to note the studies and their implications concerning eyesight.

There was an interesting statement made in Health Affairs in May of 2011: "For every 10 percent increase in high speed Internet use at the state level, associated treatment facility admissions for prescription drug abuse rose by 1 percent."[4] That may well be a good example of correlation not proving causation but it remains an interesting idea. There is certainly a strong incidence of addiction TO an electronic device. If you have ever seen someone freak out when they lost their phone, data, or contacts, you have an idea of the importance of these things to us.

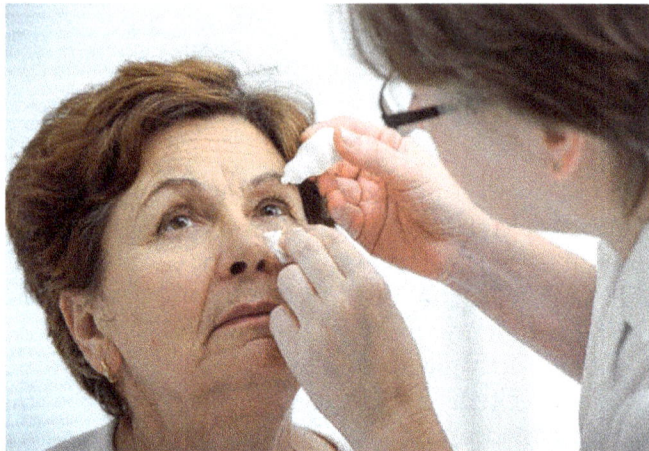

Shutterstock/Alexander Raths

Digital eyestrain is a real phenomenon. People who stare at a screen all day may be developing dry eyes, and in some cases, it may affect your near or far vision.[5] There is evidence that the "blue light" of an electronic screen could be adversely affecting your eyesight.[6] There is also an issue of an interruption of your REM sleep possibly caused by the use of an electronic screen.[7,8,9] Computer eyestrain is real.[10]

---

[3] John Biggs, "Publisher Revenues Down as Ebook Buying Slows," March 3, 2015, http://tech crunch.com/2015/03/03/publisher-revenues-down-as-ebook-buying-slows/#tmhmdj:FdUp.

[4] Anupam B. Jena and Dana P. Goldman, "Growing Internet Use May Help Explain the Rise in Prescription Drug Abuse in the United States," *Health Affairs* 30, no. 6 (May 2011), 1192–1199, http://content.healthaffairs.org/content/30/6/1192.full.

[5] "Digital Eye Strain," *The Vision Council*, accessed March 30, 2015, https://www.thevisioncoun-cil.org/content/digital-eye-strain/teens.

[6] David C., "Eye Doctors are Concerned About Blue Light (Infographic)," *VSP Blog*, July 6, 2014, http://vspblog.com/blue-light-study/.

[7] Ibid.

[8] Harvard Medical School, "Blue Light Has a Dark Side," *Harvard Health Publications*, May 1, 2012, http://www.health.harvard.edu/staying-healthy/blue-light-has-a-dark-side.

[9] Scott Sikes, "How Digital Devices are Affecting Vision," *Modern Medicine Network*, August 18, 2014, http://optometrytimes.modernmedicine.com/optometrytimes/news/how-digital-devices-are-affecting-vision?page=full.

[10] "Computer Vision Syndrome," *American Optometric Association*, 2014, http://www.aoa. org/patients-and-public/caring-for-your-vision/protecting-your-vision/computer-vision-syndrome?sso=y.

Maybe physical books will not survive. In addition, there have already been changes in how writers are remunerated. It helps, however, that you can self-publish a book. Now, if a publisher has rejected your opus, but you think it has merit, it is easier than ever to produce an electronic book for sale. There are already instances of self-published books that have gone on to "real" sales from retail outlets. It has, possibly, put a good number of editors and even publishers out of work, but it most likely is the wave of the future. It does ebb and flow, however.[11]

Shutterstock/Angela Waye

TELEVISION

Once upon a time, educational leaders truly believed that after a steady dose of television, we would all be vidiots. We would cease to think for ourselves and would spend all day plugged into a giant electronic bosom from which we would gain whatever knowledge we might eventually possess.

Many people believed that TV would replace teachers in the educational setting. Students would file in, sit in their little chairs in a row, and stare at a TV screen in order to acquire an education. A teacher would only be needed for testing. Obviously, this did not pan out as once believed. Instead, something else more interesting occurred. First, we had the abundance of electronic devices, which were far more "self-directed" learning. It was not interactive; you simply watched a screen and 'learned' in the old-fashioned sense. In addition, with the rapid growth of other types of electronic screens and access, certain demographics seemed to be abandoning the television as a medium for entertainment. In addition, those who do watch TV may be watching "canned" material (streaming or purchased DVD/movie type services). It is generally more common for people to watch streaming material on a handheld device—but this differs according to age. "Live" television has been slowly declining around the world.[12]

Shutterstock/Giacomo Pratellesi

Around the world, overall screen viewing continues to rise, but the personal handheld industry appears to be taking over, especially with younger generations of viewers. In 2002, the number of hours spent viewing television in a week in first-world countries varied from

---

[11] Adario Strange, "Nielsen Report Says Decline in Print Sales Versus E-Books Slowed in 2012," *PC Magazine*, January 9, 2013, http://www.pcmag.com/article2/0,2817,2414068,00.asp.

[12] Mark Sweney, "TV Viewing Figures Show Brits Prefer Traditional Sets Over Smartphones," *The Guardian*, February, 16, 2014, http://www.theguardian.com/tv-and-radio/2014/feb/17/tv-viewing-figures-brits-television-sets-over-smartphones.

Shutterstock/ABB Photo

Shutterstock/Lisa F. Young

28 hours (the United States and Great Britain) to 18 hours in Sweden, Norway, and Finland.[13]

There is also an interesting phenomenon: people who watch television often do so with their cell phones, laptops, or tablets in their laps. Some movie star looks familiar; you can instantly look up his prior movies—or some statement or fact can be verified. According to Nielsen (2013), people used their tablets to do general web searches (while watching TV) or web browsing. Coming in third was perusing social media sites.[14]

Think twice about borrowing someone's phone. A recent survey revealed that 75% of people with cell phones use them in the bathroom.[15]

Shutterstock/Maridav

[13] Ian Graham, "Media > Television Viewing: Countries Compared," *NationMaster*, 2003–2015, http://www.nationmaster.com/country-info/stats/Media/Television-viewing.

[14] "Action Figures: How Second Screens Are Transforming TV Viewing," *Nielsen*, June 17, 2003, http://www.nielsen.com/us/en/insights/news/2013/action-figures—how-second-screens-are-transforming-tv-viewing.html.

[15] "IT in the Toilet," *11mark*, accessed March 30, 2015, http://www.11mark.com/IT-in-the-Toilet.

Another interesting fact from the Nielsen report was that 20% of people using a tablet while watching TV were using it to shop for a product they had seen advertised.[16] This could have enormous impact on the way goods are marketed and presented on television (or any other media source). Video games as well often come with advertising of specific products included. While most of these types of statistics are essentially talking about American behaviors, the evidence seems to imply that similar behaviors occur in other first-world countries.

It is also interesting to note the rapid increase in the use of tablet technology in young children. According to Common Sense Media, 40% of families who have children aged 8 and under own a tablet.[17] The numbers of children using mobile devices has also sky rocketed. If you include video game systems, the number is astonishing. As one Time Magazine blog author (Lily Rothman) put it: "Your kids watch a full-time job's worth of TV each week."[18] While the American Academy of Pediatrics continues to recommend that a child only spend an hour or two in front of a screen each day.[19,20]

Shutterstock/wavebreakmedia

It will be interesting to see if children ever actually have their screen time reduced with the current volume of electronic devices. Perhaps such use will cause changes in brain development. In the future, such "electronic" skills might make it easier to back up brain activity and/or download new material into our brains.

CELL PHONES

Shutterstock/Monkey Business Images

There has been a good bit of research about cell phones and how they might have a negative effect on health. Since it has been shown that cell phones may "warm up"

---

[16] Ibid.

[17] "Zero to Eight: Children's Media Use in America 2013," Common Sense Media, Fall 2013, https://www.commonsensemedia.org/research/zero-to-eight-childrens-media-use-in-america-2013.

[18] Lily Rothman, "FYI, Parents: Your Kids Watch a Full-Time Job's Worth of TV Each Week," *Time Magazine*, November 20, 2013, http://entertainment.time.com/2013/11/20/fyi-parents-your-kids-watch-a-full-time-jobs-worth-of-tv-each-week/.

[19] Tia Ghose and LiveScience, "Pediatricians: No More Than 2 Hours Screen Time Daily for Kids," *Scientific American*, October 28, 2013, http://www.scientificamerican.com/article/pediatricians-no-more-than-2-hour-screen-time-kids/.

[20] American Academy of Pediatrics, "Media Use by Children Younger Than 2 Years," *Pediatrics* 128, no. 5 (2011), 1040–1046, doi: 10.1542/peds.2011–1753.

Shutterstock/Luis Louro

Shutterstock/baloon111

your brain if they are constantly attached to your head, there have been studies that imply that there is a connection between cell phones and brain cancers. The American Cancer Society has said no.[21] There have been some studies that have implied a connection (one court case was won based on the idea that a cell phone caused brain cancer[22]), but there is no good concrete scientifically based proof that there is a problem.

But wait! Scientists are looking at ways to CHARGE your cell phone using body heat![23]

It should at least be noted, however, that most data of this kind would not address the idea of children overusing cell phones. It also should be noted that cancer statistics, per se, tend to run "behind" more than most health statistics. It would be hard, for example, to say that there has been a rise in brain cancer among young people in the last five years. The cancer incidence statistics tend to be several years behind.[24]

Shutterstock/Shanta Giddens

[21] "Cellular Phones," *The American Cancer Society*, last modified December 12, 2014, http://www.cancer.org/cancer/cancercauses/othercarcinogens/athome/cellular-phones.

[22] Geoffrey Kabat, "Do Cell Phones Cause Brain Cancer? The Diehards Cling Desperately to Opinion," *Forbes Magazine*, March 5, 2013, http://www.forbes.com/sites/geoffreykabat/2013/03/05/do-cell-phones-cause-brain-cancer-the-diehards-cling-desperately-to-opinion/.

[23] Richard Gray, "Turn Yourself into a Walking Charger: Battery Harvests Energy from Body Heat and Could One Day Power Up Mobile Phones." *Daily Mail*, last updated November 26, 2014, http://www.dailymail.co.uk/sciencetech/article-2848868/Mobile-phones-charge-batteries-pocket-using-body-heat.html.

[24] Steven Novella, "New Data on Cell Phones and Cancer," *Science-Based* Medicine, May 19, 2010, https://www.sciencebasedmedicine.org/new-data-on-cell-phones-and-cancer/.

There have been some studies that implied that a male carrying a cell phone in his pocket could produce testicular abnormalities, but there is no peer-reviewed evidence to back this up. Even the World Health Organization was part of the discussion, having announced in 2011 that there was a risk involved in the use of cell phones.[25]

We are addicted to our toys. It is important to realize that they are only tools, not valuable except in that regard.

---

[25] "What are the Risks Associated With Mobile Phones and Their Base Stations?," *World Health Organization*, September 20, 2013, http://www.who.int/features/qa/30/en/.

DISCUSSION QUESTIONS

1. How much time a day do you think you spend looking at an electronic screen? Do you think it's excessive?

2. Do you think our addiction to electronic screens could be feeding our obesity issues? Why or why not?

3. Do you think that you could learn what you needed (in school) by only using an electronic device? Would you still need a (human) teacher?

4. Would you say that you are (or anyone you know is) "addicted" to social media? Why do you think so? What do you think could be done (if you think you need to) to change that?

5. Do you think that using an electronic screen before you go to bed or during the night affects your sleep? Why or why not?

6. Have you had any trouble with your eyes due to computer strain? Do you think it's a real issue? Why or why not?

RESOURCES AND REFERENCES

http://www.retale.com/info/busy-states-of-america/ (shows you in real time what someone is doing all the time)

Common Sense Media, Zero to Eight: Children's Media Use in America 2013, https://www.commonsensemedia.org/sites/default/files/research/zero-to-eight-2013.pdf.

Cell Phones and Cancer Risk, National Cancer Institute, 6/24/13, http://www.cancer.gov/cancertopics/causes-prevention/risk/radiation/cell-phones-fact-sheet.

Cancer Facts and Figures 2015, http://www.cancer.org/research/cancer factsstatistics/cancerfactsfigures2015/index.

Cancer Statistics, http://www.cancer.gov/statistics.

Religion: Introduction

"Science without religion is lame; religion without science is blind."

—Albert Einstein

What if you were a good Christian and modern science made it possible for you to live forever? What might happen to your faith in the possibility of an afterlife? One of the basic conundrums of the intersection of religion and science is that the goals of both are seldom the same. Thus, a scientific "eureka" moment might be something that disproves a strongly held religious belief.

When we talk about religion, we deal with a belief structure based upon faith. There is not always any historically accurate "proof" that someone we revere, as a religious figure, actually existed. To the faithful, it may not matter. To the scientist—someone who wants to construct an experiment to prove something—it most certainly matters. If you travel in Europe, for example, there are many, many holy relics preserved for your viewing pleasure. They are venerated objects that are often met with a degree of skepticism (how many fingers did that Saint have?). Let us assume that we could actually do scientific tests on these venerated bones or body parts. Possibly we could then prove unquestionably that they really are as claimed (assuming that we had access to the DNA of an ancestor, for example). A scientist might then announce this to the world as an amazing discovery. Would it change the belief structure of the faithful? Probably not.

A thousand years ago, an ancestor saw an eclipse and assumed it was some sort of magic or an act of God that had no explanation. Today, we can explain such a phenomenon in scientific terms. Can we then "prove" that everything can now be understood? Again, probably not.

There is probably little point to attempting to change someone's belief structure using the scientific method. It isn't science that makes you believe. Unfortunately, this dichotomy between science and religion is causing even more polarity when it comes to behavior.

Religion: Religion in the Electronic Age

"Flight from and hatred of technology is self-defeating. The Buddha, the Godhead, resides quite comfortably in the circuits of a digital computer or the gears of a cycle transmission as he does at the top of a mountain or in the petals of a flower. To think otherwise is to demean the Buddha – which is to demean oneself."

—Robert M Pirsig (Zen and the Art of Motorcycle Maintenance)

Sometimes it is hard to reconcile the religions of earlier centuries with the variety of religions we have today. By the same token, much is the same. During the Crusades, the rhetoric was presented as a conflict between religious faiths; the reality was more likely based upon economic concerns. Religion was used as a helpful crutch to acquire (or not) what someone wanted.

Today, the Internet and all sorts of visual media make this quicker and more dangerous. We can watch a video of someone being beheaded for religious differences; we can view unrest based upon religious ideals. In one sense, this tends to split factions apart even more. If you view someone who is dressed differently, speaks a different language, believes in different things, you feel an even larger degree of separation.

Scientists would have you believe that science can be a more level playing field—that science can explain something without religious belief involved. I suspect that the famous Islamic scientists (called Saracens at the time) during the Crusades might disagree. While one religion might say "this is about getting area X back to the true believers," there were always multiple factors involved. Perhaps we might get a nice spice trade and some money as well. Neither

Shutterstock/Feng Yu

Shutterstock/Muemoon

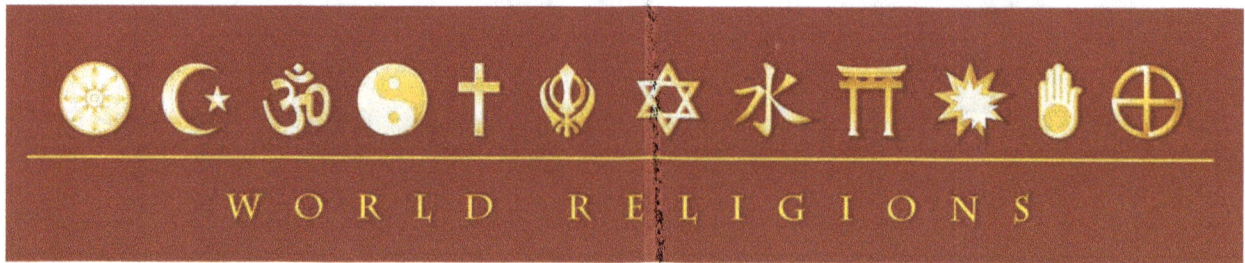

WORLD RELIGIONS

Shutterstock/casejustin

science nor religion could ever be the only cause for any conflict. Either might have been a starting point, but other factors were always involved.

A discussion of the many and varied world religions is beyond the scope of this textbook, but it is important to note how various religious belief structures might affect the development of science and technology. Before the Industrial Revolution, people generally trusted religion over science; if you saw a "sign" in the natural environment, many people tended to ascribe it to a religion or a godlike being. The "order" of the universe was ascribed to a godlike figure. Scientific thinkers were the radicals of the time. To ascribe an event to a scientific cause or reason was anathema.

Shutterstock/KeilaNeokow EliVokounova

There was a certain correlation between believing in science and being "modern." People began to slowly put more faith into the idea that there were answers and that something might be controllable. Imagine seeing someone having a seizure (in historic times, that might be ascribed to being possessed by demons). Once treatment or medication entered the picture, people began to have more faith in the scientific approach.

In some respects, the idea of simply accepting "God's will" was overtaken by the idea that you should help yourself. Science was the "new" way of thinking. Every problem had a solution. You were no longer trapped in a way of thinking that prevented you from ever overcoming something.

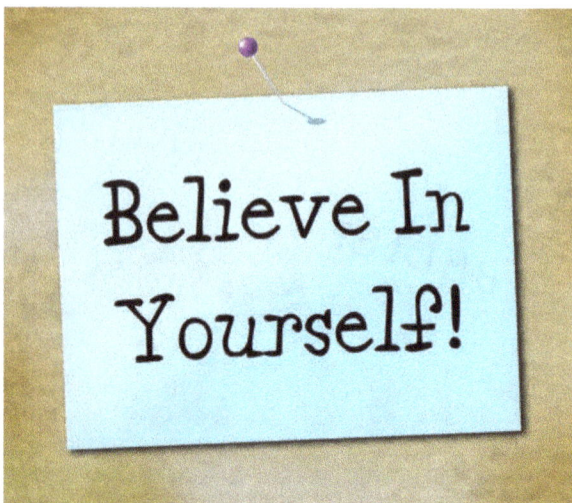

Shutterstock/Stuart Miles

If you look at the movie, Twenty Thousand Leagues Under the Sea, you see a main character that represents the new head of the food chain. No longer was a religious institution at the top (or a political government). Captain Nemo was an individual and a scientist. He had no religious belief; he was not hampered by any nationalistic feelings. Connected to no place, belief, or government, he represented what some though might become the new world order. At various points in history, we have looked at inventions and proclaimed them the "answer" to whatever societal ills existed. Electricity was going to solve poverty. Atomic energy would solve any energy crisis. The "right" world order was one where science simply designed a solution to whatever problem occurred.

Shutterstock/mtkang

Shutterstock/Neftali

You did not need anything else. In some respects, religion was shuffled aside as being the "old" way of thinking.

Moving forward 200 years, we now have a robust scientific tradition that, in many ways, has changed how we view the world. Unfortunately, science also has some issues with the display of information and the belief structure we ascribe to it. Today, the chasm between organized religion and science still exists.

Both sides have suffered from the rise of the Internet. Religious dogma can stifle scientific endeavors with publicity about a discovery that is not in keeping with certain religious views. By the same token, scientists can publish a small experiment proving "X" while later experiments show that X is not correct. There is also the aspect of telling something online that could be extremely dangerous. If you discover a bioweapon by accident, do you "tell all" so that it could be conceivably so used on you and yours?

Today, a religious figure can far more easily display something on a television screen and claim that it is a miraculous healing. While miracle healings might well occur, seeing them on television is a bit problematic since such images are so easily manipulated. There are certainly as many pseudoscientific items on the screen as well. In this day and age, it is easy to manipulate an audience by fake results.

While the numbers for organized religion might be slowly decreasing, religion remains a powerful, organized structure throughout the world. According to the Pew Charitable Trust, in 2010, 31.5% of the people in the world were Christians, 23.2% were Muslims, 15% were Hindus, and 7.1% were Buddhists.[1] The study also measured the degree of religious diversity in each area of the world. In the United States, 78% of the people

Shutterstock/Marilyn Volan

[1] "Global Religious Diversity: Half of the Most Religiously Diverse Countries Are in Asia-Pacific Region," Pew Research Center, April 4, 2014, http://www.pewforum.org/2014/04/04/global-religious-diversity/.

were listed as Christians; in Iran, over 99% were listed as Muslims.[2]

In one sense, the rise of science has caused even more polarity. If there are scientific concepts that you honestly believe go against the tenets of your faith, then that may be more polarizing. Religious opposition to certain beliefs (like diversity) is beginning to drive political structures, for example.[3]

It will also be curious to see if the rise of social media will affect religious behavior. Some churches and synagogues are embracing new technology—you can send a tweet to the priest or pastor or rabbi while he or she is talking if you don't understand something. You can project images during a service to expand or clarify a point. According to the Pew Research Center, one in five Americans talks about their faith on social media platforms.[4] It is nearly routine to see, in certain American demographics, a request for prayer on social media sites.

The darker side of religion and the Internet is that it broadens the "appeal" base for certain kinds of scams. If you have a certain belief structure and someone online requests money for further action in that area, it is far easier to be misled. If you believe that X religious person is "godly," then you might respond when you see online that he or she needs money to fight the "ungodly." Of course, such scams have always existed and preyed on many people. The Internet just provides a bigger fan base.

Shutterstock/Elena Schweitzer

Shutterstock/KieferPix

---

[2] Ibid.

[3] "Public Sees Religion's Influence Waning," *Pew Research Center*, September 22, 2014, http://www.pewforum.org/2014/09/22/public-sees-religions-influence-waning-2/.

[4] Alan Cooperman, Greg Smith, Becka Alper, and Katherine Ritchey, "Religion and Electronic Media: One-in-Five Americans Share Their Faith Online," *Pew Research Center*, November 6, 2014, http://www.pewforum.org/files/2014/11/Religion-and-Electronic-media-11-06-full.pdf.

DISCUSSION QUESTIONS

1. What do you think is the difference between myth and religion?

2. Do you think that ultimately science will "prove" that religion is not based on fact? Will that matter to people?

3. Do you think the world needs religion and science both? Why/Why not?

4. Do you think that religion is "robust" enough to change with new scientific discoveries?

5. Do you think that man is the only animal with a soul (and thus capable of religion)? Why/Why not?

6. If you had access to a "proof" that would deny the existence of religion or faith, would you feel honor-bound to tell everyone? Why?

7. Would you rather live in a world without religion...or a world without science? Do you think that science will disprove God or God will outlast science? Why/ Why not?

RESOURCES AND REFERENCES

Pew Research Center, "The future of World Religions: Population Growth Projections, 201–2050" 4/2/2015 http://www.pewforum.org/2015/04/02/religious-projections-2010-2050/

MIT Technology Review, "How the Internet Is Taking Away America's Religion." 4/4/2014 http://www.technologyreview.com/view/526111/how-the-internet-is-taking-away-americas-religion/

Website: Society, Religion and Technology http://www.srtp.org.uk

Religion: Miracles

"Miracles are not contrary to nature, but only contrary to what we know about nature."

—St. Augustine

If you were to question a group of people about the subject of miracles, most people will generally admit that miracles might really occur. Americans, especially, tend to believe that miracles do occur (although the number is waning). In 2013, according to a Harris poll, 72% of the Americans polled stated that they believed in miracles. In addition, millennials are more likely to believe in miracles (80%). Strangely enough, those numbers are not matched by religious belief. Not as many people believe in God or attend church regularly.[1] While it is difficult to divorce the idea of a miracle from a specific religious ideology, it is interesting to consider how a modern scientist might perceive the idea of a miracle.

Long ago, a miracle was often a phenomenon that we could not explain. An eclipse might be viewed in that light if you knew nothing about the rotation of heavenly bodies. To prehistoric man, nearly everything that happened could be so construed. The first fire might have inspired the same awe as a religious miracle of later years. In modern times, a miracle would have to be defined as something that was impossible in nature, thus a "supernatural" event. Something must happen that beats all the odds that it could occur. It is not something that man can (given what we know) actually do or replicate. Generally speaking, it has to be "public" and viewable. You should be able to empirically verify that it occurred.

Many different religions believe in miracles. Some, like the Muslims, do not believe that such things should be dissected or discussed; they are acts of faith. Christians also believe in divine intervention, as a rule, when it comes to miraculous occurrences. In the New Testament, Jesus was said to perform many miracles, setting the stage for this belief. Thomas

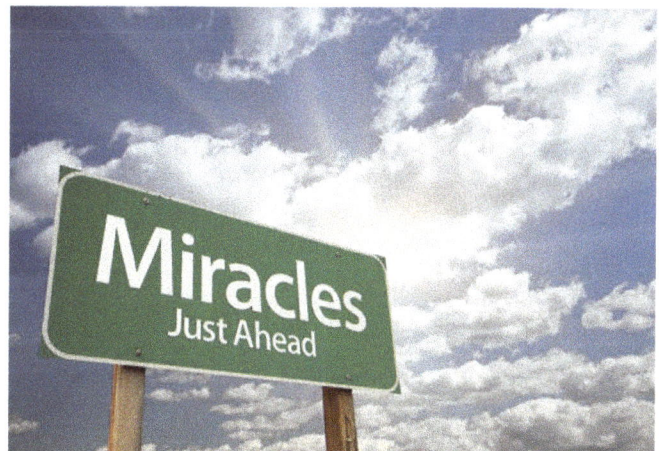

Shutterstock/Andy Dean Photography

---

[1] Larry Shannon-Missal, "Americans' Belief in God, Miracles and Heaven Declines: Belief in Darwin's Theory of Evolution Rises," *Harris Polls*, December 16, 2013, http://www.harrisinteractive.com/NewsRoom/HarrisPolls/tabid/447/ctl/ReadCustom%20Default/mid/1508/ArticleId/1353/Default.aspx.

Shutterstock/Celiafoto

Jefferson, for one, did not believe in the miracles in the Bible. He saw them as outside the laws of science and thus invalid. He took his New Testament and physically cut out all of the miracles. His Jefferson Bible is still available today.

The Catholic Church, however, primarily because of sanctification practices, has an extensive process by which miracles are validated. The rules (established in 1983) include two stages. In the diocese where the miracle occurred, the local bishop has to open an enquiry and question witnesses. All documentation about the case must be collected before being sent on. After that, the "Roman Congregation" must pass judgment on whether or not a miracle actually occurred. There are basically three categories: resurrection from the dead, the curing of an incurable disease (including organs or bones growing back), and curing a disease immediately that normally would take a long time.[2]

For a miracle to be "approved," it is first given to the Consulta Medica—a group of Catholic physicians who review all of the medical information. The majority of cases are judged to not be miracles.[3] The point here is that there is an extensive "vetting" process for the occurrence of a miracle. Less than a hundred such instances have been "proven" to be miracles. Science has no answer (or, as the scientists would say, not yet).

Shutterstock/marekusz

Each of the major religions recognizes that miracles occur, but all are generally attributable to a non-earthly cause. How modern science and modern religion look at miracles provides good information on how science and religion intersect. There are still many things in this world that we cannot simply explain away with science. Perhaps the scientists are right and someday we may be able to.

---

[2] "Congregation for the Causes of Saints," *The Vatican*, accessed March 28, 2015, http://www.vatican.va/roman_curia/congregations/csaints/index.htm.

[3] John Collins Harvey, "The Role of the Physician in Certifying Miracles in the Canonization Process of the Catholic Church III," *Southern Medical Journal* 100, no. 12, December 2007, 1255–1258, https://pmr.uchicago.edu/sites/pmr.uchicago.edu/files/uploads/Harvey%2C%20The%20Role%20of%20the%20Physician%20in%20Certifying%20Miracles%20in%20the%20Canonization%20Process%20of%20the%20Catholic%20Church%20III_0.pdf.

DISCUSSION QUESTIONS

1. Do you think miracles really happen? Why?
2. Do you think that everything will one day be explainable by science? Why or why not?
3. Have you experienced anything you would consider miraculous in your own life or in your family? Explain.
4. If there is no validity to religion, then how might you explain something that is outside the "rules" of nature?

RESOURCES AND REFERENCES

Sulmazy, Daniel P., What is a miracle?, *Southern Medical Journal* • Volume 100, Number 12, December 2007, 1223, http://inters.org/files/What-is-a-miracle.pdf.

Website: www.miraclehunter.com.

Haggerty, Barbara Bradley. A boy, an injury, a recovery, a miracle?, NPR, http://www.npr.org/2011/04/22/135121360/a-boy-an-injury-a-recovery-a-miracle.

Do You Believe in Miracles? Most Americans Do. NPR, Talk of the Nation, 2/25/2010, http://www.npr.org/templates/story/story.php?storyId=124007551.

Religion: Scientific Proof and Belief

"Technos and clerics have much in common. Both take a world that can't be fully understood and try to explain its fundamental properties."

—Dan Ronco

In Cobb County, Georgia in 2002, new science books came with a sticker affixed to the front cover. The sticker read:

"This textbook contains material on evolution. Evolution is a theory, not a fact, regarding the origin of living things. This material should be approached with an open mind, studied carefully, and critically considered."

The creationists and the intelligent design people in the county were offended at the teaching of evolution in the textbooks. The stickers brought news and notoriety to the county. After a lengthy court case, *Selman v. Cobb County School District*, 449 F.3d 1320 (11th Cir. 2006), the sticker was finally removed. This is a good example of the wealth and abundance of scientific knowledge losing (originally) against a small group of religious people. They did not want evolution taught as the only possibility. It didn't matter to them that the majority of scientific experts agreed with the concept of evolution. It was not the only court case for similar reasons.[1,2] The scientific community also had its share of court cases. In *Tammy Kitzmiller, et al. v. Dover Area School District, et al.* (400 F. Supp. 2d 707, Docket no. 4cv2688), suit was brought that wanted to eliminate the teaching of intelligent design in public schools. The court ruled that the teaching of intelligent design was essentially a religious concept and thus should not be taught.[3]

Shutterstock/livinglegend

---

[1] Laura Parker, "School Science Debate Has Evolved," *USA Today*, November 28, 2004, http://usatoday30.usatoday.com/news/education/2004-11-28-schools-evolution_x.htm.

[2] *Selman v. Cobb County School District*, 449 F.3d 1320 (11th Cir. 2006), *Justia US Law*, May 25, 2006, http://law.justia.com/cases/federal/appellate-courts/F3/449/1320/521071/.

[3] "Kitzmiller v. Dover: Intelligent Design on Trial," *National Center for Science Education*, October 17, 2008, http://ncse.com/creationism/legal/intelligent-design-trial-kitzmiller-v-dover.

Shutterstock/Filipe Frazao

There have always been scientific ideas (and even real happenings) that were met with opposition from the religious sector. From the conspiracy theorists believing that we never went to the moon (it was all faked in a TV studio) to the more recent deniers of climate change, science can still be (if only temporarily) stopped or stymied by strongly held beliefs. Generally speaking, scientists are less likely to be religious than non-scientists.[4] Many people were upset when President Obama named Francis Collins to be the new head of the National Institutes of Health. Dr. Collins had a strong religious background and some saw it as evidence that he would not be as "scientific" as he needed to be.

The concept of scientific proof does not apply to a religious belief. Scientists (like many rational people) believe that if they simply explain things clearly (sometimes a problem), then people will understand and agree. As we discussed earlier, the Dunning-Kruger effect can prevent someone from understanding even in the face of overwhelming proof.

The argument is the most vociferous when it comes to education. People with strongly held religious beliefs would like their beliefs to be taught to their children (thus, so many home school). Society has moved into a position that is leaning more and more toward accommodation to all. We no longer pray in school; some religious organizations cannot function inside a public school because of the fear that some 'other' group would demand equal time.

Shutterstock/Andresr

The denial of scientific concepts is not universal by all religions. In 2015, the presiding Bishop of the Episcopal Church, Katherine Jefferts-Schori, stated that to deny climate change was immoral—she feels strongly that God gave you a brain and you should use.[5] In December of 2014, it was announced that Pope Francis planned to issue an edict on the problem of climate change with a plea for all Catholics to protect the environment.[6]

When it comes to technology, however, one very interesting religious community can help us to understand some of the critical problems between religion and technology. Most

[4] "Scientists and Belief," *The Pew Charitable Trusts*, November 5, 2009, http://www.pewtrusts.org/en/research-and-analysis/reports/2009/11/05/scientists-and-belief.

[5] Suzanne Goldenberg, "Climate Denial Is Immoral, Says Head of US Episcopal Church, *The Guardian*, March 24, 2015, http://www.theguardian.com/environment/2015/mar/24/climate-change-denial-immoral-says-head-episcopal-church.

[6] Nicholas St. Fleur, "What Can a Popular Pope Do About Climate Change?", *The Atlantic*, December 30, 2014, http://www.theatlantic.com/international/archive/2014/12/what-can-a-popular-pope-do-about-climate-change/384119/.

people believe that the Amish community does not believe in the use of technology. Seeing horse-drawn buggies in Amish country is pretty common. The Amish, however, actually have a different "test" for the use of any tool or implement (from an axe to a cell phone). You can see an Amish person using a public phone, or borrowing one from a neighbor. You might think that it is someone who is "breaking the rules," but the actual test for use is far more interesting and complex.

Shutterstock/Vladislav Gajic

Regulations differ from area to area but the decision for an Amish community to use a certain item is based upon the idea of whether it actually follows or enhances the basic beliefs of the community. Using some technology might make you prideful (thus losing your humility). It's a link to the outside world where different values prevail. An Amish business might not have a computer, but they might have a website to sell their materials. The decision of use is based upon the concept that the tool (of any description) must not be any intrusion into family life. It must enhance and strengthen the community values.[7] They are not, as many believe, anti-technology. One of the reasons that they are opposed to cars is that they feel that a car breaks down community ties.[8]

The difference lies in their deliberation before the use of any technology. It is also not a monolithic community; each small community can make a decision for the good of the people there. It is not a religion wedded to old ideas; in some ways, it is very modern. Every tool is evaluated. Will it benefit the community? Sometimes, a tool will be accepted and later rejected if it appears to hurt the community structure in any way. This kind of test might be a valid litmus paper for any use of technology.

---

[7] "The Pennsylvania Amish & Their Use of Technology," *Discover Lancaster*, accessed March 30, 2015, http://www.discoverlancaster.com/towns-and-heritage/amish-country/amishandtechnology.asp.

[8] "Amish Tech," *Amish Country News*, accessed March 30, 2015, http://www.amishnews.com/amisharticles/amish_tech.htm.

DISCUSSION QUESTION

1. What do you think of the Amish test of technology?
2. Do you think that belief concepts that are essentially non-scientific should also be taught in public schools? Why or why not?
3. Do you think it's "fair" for a school to forbid religious activity like prayer?
4. Do you think certain religious clothing rules should be legal in public school?

RESOURCES AND REFERENCES

Brady, Jess. Amish Community Not Anti-Technology, Just More Thoughtful. NPR, 9/2/2013, http://www.npr.org/blogs/alltechconsidered/2013/09/02/217287028/amish-community-not-anti-technology-just-more-thoughful

Kraybill, Donald B., Johsnon-Weiner, and Nolt, Steven M. *The Amish.* Baltimore: Johsn Hopkins University Press, 2001.

All Things Considered Radio Story, http://www.npr.org/blogs/alltechconsidered/2013/09/02/217287028/amish-community-not-anti-technology-just-more-thoughful (also cited below)

The Amish, The American Experience, PBS, http://www.pbs.org/wgbh/americanexperience/features/introduction/amish-introduction/

Technology and Robotics: Introduction

"R2-D2, you know better than to trust a strange computer."

—C3PO, Star Wars Episode V

ARTIFICIAL INTELLIGENCE

Artificial intelligence is not just computer science. The concept also incorporated physiology and philosophy. There is no truly clear definition, which is part of the problem. If you ask a group of students (or computer scientists) to define AI, you will get a variety of responses. Mechanically speaking, it's the idea of developing a machine that functions like a brain—but even that definition is open to interpretation. We are pretty clear on the "machine" part, but defining "intelligence" has been a somewhat fluid concept for centuries. Once upon a time, humans designed intelligence quotient tests (IQ) in order to measure how smart someone was. Somewhat later we realized that perhaps that did not measure the true impact of intelligence.

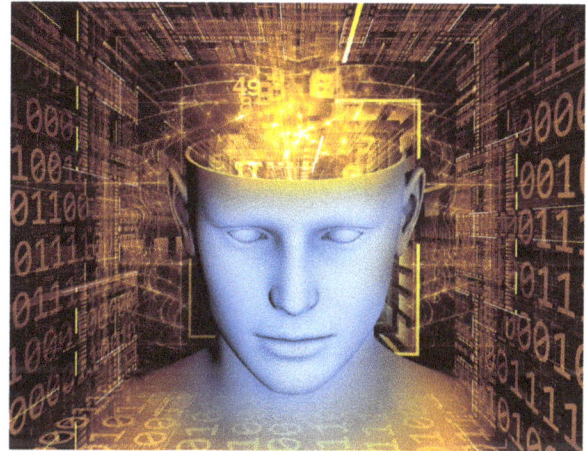

Shutterstock/agsandrew

A computer was originally designed to calculate things based upon a rapid sequence of ones and zeros which could also be construed to be yes or no. Philosophically, along the way, we came to separate the brain from the body. If we could separate mental from the physical attributes, that might make it easier to quantify "intelligence." Think of it in terms of how we study heat or light versus how we get burned—reason versus experience. That might seem overly simply, but the difference is enormous for a computer "brain."

In the early 19th century, Charles Babbage wanted to design a machine that would do mathematical computations. The idea was that it could eliminate both mistakes and "drudgery" for human beings. The idea that we could design a machine to do such chores was somewhat radical; it also helped to engender the idea that the machine was "thinking" when it fact, it was only doing a mechanical process more quickly than a human

"On two occasions I have been asked,—'Pray, Mr. Babbage, if you put into the machine wrong figures, will the right answers come out?' . . . I am not able rightly to apprehend the kind of confusion of ideas that could provoke such a question."[1]

---

[1] "Charles Babbage," *wikiquote.org*, last modified March 27, 2015, http://en.wikiquote.org/wiki/Charles_Babbage.

Alan Turing 1912–1954
Mathematician and WWII code breaker

Shutterstock/Neftali

could. It was a separation of the thing done from the idea about the thing being done.

If we start with the idea that "learning" can be quantified into a series of yes or no questions, then we can understand the development of the computer. In 1950, Alan Turing designed the famous "Turing Test" (also called the Imitation Game). It gave us a possible model for future AI as well as the possible implausibility of actually producing artificial intelligence. It didn't try to answer the "hard" questions, its premise is based upon the idea that what we know could be broken down into a series of empirical questions in order to "find" the right answer.

It basically takes human interaction and places a computer in the middle. A machine is one room and a human in another. There is a second human (called an "interrogator" who cannot see either the other human or the computer. The interrogator must decide where the human is by simply asking a list of questions. While you might immediately want to deny that this is "true" intelligence or learning, it should be pointed out that this is the most common way we currently have to decide if someone is "smart" or not—we ask questions.

Turing realized that there would be two ways to build a smart machine—from the top down (by explicit programming) or from the bottom up (learn as a child learns).[2] In 2014, it was alleged that someone had "beat" the Turing Test[3] while others refuted the "proof"[4] which only proves that the issue has yet to be resolved.

Shutterstock/Mclek

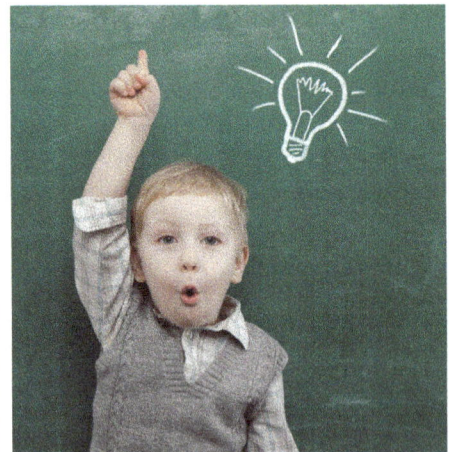

Shutterstock/YuryImaging

[2] Andrew Hodges, "The Turing Test, 1950," *The Alan Turing Internet Scrapbook*, accessed April 1, 2015, http://www.turing.org.uk/scrapbook/test.html.

[3] Terrence McCoy, "A Computer Just Passed the Turing Test in Landmark Trial," *washingtonpost.com*, last modified June 9, 2014, http://www.washingtonpost.com/news/morning-mix/wp/2014/06/09/a-computer-just-passed-the-turing-test-in-landmark-trial/.

[4] Mike Masnick, "No, a 'Supercomputer' Did Not Pass the Turing Test for the First Time and Everyone Should Know Better," *techdirt.com*, last modified June 9, 2014, https://www.techdirt.com/articles/20140609/07284327524/no-supercomputer-did-not-pass-turing-test-first-time-everyone-should-know-better.shtml.

The actual phrase, artificial intelligence, was allegedly coined by John McCarthy for a Dartmouth University in 1955.[5]

If you think about the performance done by the IBM computer known as "Watson" on television game shows, you can see a recent by-product of this kind of thinking. We tend to learn by incremental steps, and that is how Watson "learned" what it needed to know. This is only one method of artificial intelligence. It is very difficult to "prove" that this is brain-like functioning, but it does match some concepts of learning if not intelligence.

Other methods of artificial intelligence are also being developed. Aside from the "learning" machine approach, we are also looking at biological models of neural networks. If the brain is chemistry, then perhaps we can physically create a matching model. It is difficult to say if we are now (or ever will be) "close"—especially in light of the fact that we do not understand all we need to about intelligence.

Shutterstock/Everett Collection

People who can sit and discuss heavy philosophical ideas are not always useful when it comes to cutting wood. On the other hand, a machine that uses complex algorithms to decide just what we might need to see as news can now replace the old idea of a cigar chomping old guy deciding what was "news."

---

[5] John McCarthy, Marvin Minsky, Nathan Rochester, and Claude Shannon, "A Proposal for the Dartmouth Summer Research Project on Artificial Intelligence," *Stanford.edu*, August 31, 1955, http://www-formal.stanford.edu/jmc/history/dartmouth/dartmouth.html.

DISCUSSION QUESTIONS

1. Which are you more comfortable with—an old guy deciding, "if it bleeds it leads" or a computer program—for getting news?
2. How would you define artificial intelligence?
3. Do you think we have AI now?
4. Do you think computers like Watson are displaying artificial intelligence?

RESOURCES AND REFERENCES

www.computerhistory.org.

Turing, A. M., Computing Machinery and Intelligence, http://www.abelard.org/turpap/turpap.php.

www.turing.org/uk.

Association for the Advancement of Artificial Intelligence, http://www.aaai.org/home.html.

Kelly, Kevin, The Three Breakthroughs That Have Finally Unleashed AI on the World, Wired, 10/27/14, http://www.wired.com/2014/10/future-of-artificial-intelligence/.

Robots Learning Like Humans | Through the Wormhole with Morgan Freeman, Science Channel https://www.youtube.com/watch?v=48Fh25bXvqk.

A real debate on artificial intelligence, Mark Ronson gets lyrical, plus a look at why diversity is important in tech, Ted Talk, 1/22/15 http://blog.ted.com/ted-community-news-1-22-15/.

Technology and Robotics: The Laws of Robotics

"Man is a robot with defects."

—Emile Cioran, The Trouble With Being Born

"Let's start with the three fundamental Rules of Robotics. . . We have: one, a robot may not injure a human being, or, through inaction, allow a human being to come to harm. Two, a robot must obey the orders given it by human beings except where such orders would conflict with the First Law. And three, a robot must protect its own existence as long as such protection does not conflict with the First or Second Laws."

ISAAC ASIMOV, Astounding Science Fiction, Mar. 1942

In a classroom today, many students are familiar with Isaac Asimov's "Three Laws of Robotics." When asked, most can cite the general gist of the idea. The problem is that many people think these are really "rules" although they are hard put to define just who might have produced (or monitored) such "rules." The idea that a robot should not be able to hurt you is a lovely concept, but it bares no relation to any degree of law or proof. Asimov, himself, gave us the word, "Robotics."

The word Robot was first used by Karel Capek in a 1921 play called, "R.U.R." which stood for "Rossom's Universal Robots." He gave his brother, Josef credit for the name, which literally means some sort of serf labor. If the play we remotely predictive of the future, there is trouble ahead. The robots of the play try to take over the world.[1] Capek died in 1938.[2] Popular in the public imagination (and science fiction), the idea of a robot began as a fascination, but soon came to be the stuff of horror movies.

Generally speaking, the modern development of robotics revolved around the "dull, dirty, and dangerous" idea. If we are going to build

Shutterstock/josefkubes

---

[1] Dominik Zunt, "Who Did Actually Invent the Word 'Robot' and What Does It Mean?," accessed April 1, 2015, http://capek.misto.cz/english/robot.html.

[2] Ibid.

Shutterstock/soo hee kim

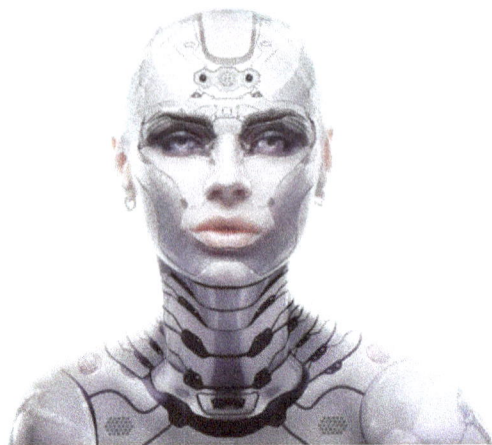

Shutterstock/Willyam Bradberry

robots, we need to figure out why we need them; otherwise, they might never get the cash infusion to produce working examples. This DDD concept has come to also have a "boring" aspect; that is, we want robots to do what we are too lazy to do in our modern lives.

Once we have figured out just what we might want robots for, there is an important question to ask. Do we want this robot to look human or not? On the "face" of it, you might have an immediate response. Depending on your age, you might think that a mechanical looking toy is cool. If you are much older or much younger, human appearance might be more important. Imagine, for example, if you needed childcare for an infant. Do you want a caretaker that looks more like Mrs. Doubtfire or Wall-E? While both might have the same attributes, you might want a more human looking caretaker. Perhaps in the future, this might not be true (once we are more accustomed to looking at robots). How about smell? Would you want it to have a more comforting "human" smell or would you prefer the smell of mechanical parts?

If you had the animal as a pet, wouldn't you want it to look more "real" (especially if you were a child who had a robot animal because you were allergic to the real thing)? There is currently a "therapeutic" robot named Paro, which is used in nursing homes and medical situations in order to provide comfort.[3]

The primary use of robots today, however, is in industrial situations. Robots have replaced humans in many large-scale industrial applications. According to the International Federation of Robotics, there were 178,132

Shutterstock/Julien Tromeur

Shutterstock/Nataliya Hora

[3] "PARO Therapeutic Robot," *parorobots.com*, accessed April 1, 2015, http://www.parorobots.com.

units of robots sold in 2013[4] and approximately 225,000 units sold in 2014.[5] The automotive industry was the largest user of industrial robots.[6] They also stated that by the end of 2013, there was "between 1,332,000 and 1,600,000 units in operation.[7]

The actual use of robots, however, has no limit. There are examples of classrooms with robot teachers. In South Korea, for example, there is a shortage of English teachers that has been met by the use of humanoid robots.[8] This can be a physical humanoid robot or a "telepresence" type robot.[9] Robots in Japan can more easily teach the intricacies of writing to children.[10]

There is also interesting work being done with robots and autistic children. For example, a child with Asperger's Syndrome (a high-functioning type of autism) can be "socialized" by using a humanoid robot. There is no "cure" for autism, but a child can be taught to recognize social signals with the use of a robot. The robot (which is usually considered to be nonthreatening) can mimic happy or sad and label it for the child. By consistent patterning, a child can "learn" appropriate social behaviors.

By interacting with a robot "child," the autistic child can learn about both expressions and social interactions in a safe and nonthreatening environment. It is not inconceivable that an ill child could control a humanoid double sitting in a classroom.

Shutterstock/Dejan Stankovic

Shutterstock/Photodiem

---

[4] "Industrial Robotics Statistics," *International Federation of Robotics*, accessed April 1, 2015, http://www.ifr.org/industrial-robots/statistics/.

[5] "Global Robotics Survey: Industrial Robot Sales go Through the Roof Exceeding Demand from Asia," *International Federation of Robotics*, accessed April 1, 2015, http://www.ifr.org/news/ifr-press-release/global-robotics-survey-703/.

[6] Ibid.

[7] "Industrial Robotics Statistics," *International Federation of Robotics*, accessed April 1, 2015, http://www.ifr.org/industrial-robots/statistics/.

[8] Susannah Palk, "Robot Teachers Invade South Korean Classrooms," *cnn.com*, last modified October 22, 2010, http://www.cnn.com/2010/TECH/innovation/10/22/south.korea.robot.teachers/.

[9] Michael Powell, "Robot Teachers in the Classroom," *iQ Intel Magazine*, last modified October 30, 2014, http://iq.intel.com/robot-teachers-in-the-classroom/.

[10] Agence France Presse, "Calligraphy Robot Teaches Japan's Schoolchildren the Art of 'Shodo' Writing," *The World Post*, last modified August 1, 2013, http://www.huffingtonpost.com/2013/08/01/calligraphy-robot-japan_n_3686261.html.

Shutterstock/KPG_Payless

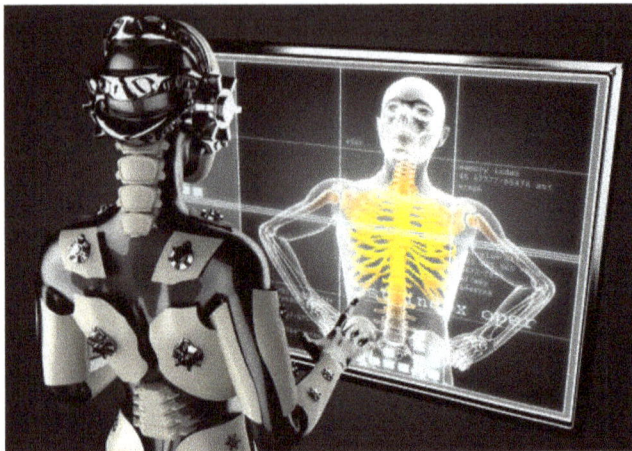

Shutterstock/videodoctor

Julien Tromeur

Robots are also becoming extremely valuable in disaster aid and assistance. Without the issues of fatigue and personal safety, a robot can search indefinitely and in places that people could not. There have been prototype lifeguard robots, for example, that save people in the ocean.[11] While the cost is currently still prohibitive, it is easy to see that we might one day have robots in jobs where there are no workers.

A good example of this might soon be the nursing field. We already have robot surgical systems like the da Vinci surgical System. Google already owns several robotics companies (like Boston Dynamic and the Google X Life Sciences division) that are actively working on robotic surgery and other medical applications.[12]

Although the idea of a robot nurse can be a bit upsetting, it might be a far safer alternative (especially for the safety of the caretaker). A robot would not have weight constraints or run the risk of illness. In addition, there continues to be a shortage of healthcare workers throughout the world and this type of technology might provide a valid answer. We already have computers doing a very effective job of pathology reports and other types of medical testing. A robot could scan through millions of possibilities when it comes to diagnosis as well.

Some countries, like Korea, have moved to using robots as prison guards. The efficacy is immediately apparent. They can't be bribed or physically hurt and they can't take sick leave and so on.[13]

The varieties of ways in which we might use robotics is enormous. At this point, it is important to realize why we don't use them even more than we do. First and foremost, is cost effectiveness. But, second to that, and equally as important is the relationship between the humans and the machines.

11 "Robot Lifeguard 'Emily' Invented to Save Swimmers," The Huffington Post, last modified May 29, 2011, http://www.huffingtonpost.com/2011/03/29/robot-lifeguard-emily_n_842304.html.

12 Tim Moynihan, "Google Takes on the Challenge of Makind Surgery Safer," *Wired*, last modified March 30, 2015, http://www.wired.com/2015/03/google-robot-surgery/.

13 Lena Kim, "Meet South Korea's New Robotic Prison Guards," *Digital Trends*, last modified April 21, 2012, http://www.digitaltrends.com/cool-tech/meet-south-koreas-new-robotic-prison-guards/.

DISCUSSION QUESTIONS

1. If you could have a robot in your home, what would you want it do?
2. Would you want the robot (in your home) to look human or not?
3. Would you want the robot to be programmed with human emotions?
4. If it looked human and broke down, would you easily be able to destroy or replace it?
5. Would you trust a robot doctor or surgeon?

RESOURCES AND REFERENCES

http://www.dulldirtydangerous.com.

R. U. R. (play by Karel Capek), https://ebooks.adelaide.edu.au/c/capek/karel/rur/act3.html.

International Federation of Robotics, http://www.ifr.org/industrial-robots/statistics/.

Robotic Industries Association, http://www.robotics.org/Industry-Statistics.

Cheng, Maria, Robot Helps Autistic Kids, The Huffington Post, 3/9/11, http://www.huffingtonpost.com/2011/03/09/robot-helps-autistic-kids_n_833321.html.

Carey, Benedict and Markoff, John, "Students, Meet Your New Teacher, Mr. Robot," NY Times, 7/10/2010, http://www.nytimes.com/2010/07/11/science/11robots.html?_r=0.

Griffiths, Andrew, The robot teacher connecting with autistic children, The Telegraph, 2/14/2014, http://www.telegraph.co.uk/technology/news/10632937/The-robot-teacher-connecting-with-autistic-children.html.

da Vinci Surgical System, www.davincisurgery.com.

Robert Full, Ted Talk, "The Secrets of nature's grossest creatures, channeled into robots." 3/2014 http://www.ted.com/talks/robert_full_the_secrets_of_nature_s_grossest_creatures_channeled_into_robots

Dennis Hong, Ted Talk, "My seven species of robot, 9/2009 http://www.ted.com/talks/dennis_hong_my_seven_species_of_robot.

P. W. Singer, Ted Talk, Military robots and the future of war, 2/2009 http://www.ted.com/talks/pw_singer_on_robots_of_war.

Robot prison guard in South Korea https://www.youtube.com/watch?v=stgZ1nFWEmE.

Goldberg, Ken, Ted Talk, 4 lessons from robots about being human, 2/2012 http://www.ted.com/talks/ken_goldberg_4_lessons_from_robots_about_being_human#t-439077.

Humanoid Robot—Gemonoid HI-1 https://www.youtube.com/watch?v=uD1CdjlrTBM.

Technology and Robotics: Why Robotics?

Detective Del Spooner: "Human beings have dreams. Even dogs have dreams, but not you, you are just a machine. An imitation of life. Can a robot write a symphony? Can a robot turn a. . .canvas into a beautiful masterpiece? Sonny (robot): "Can you?"

—iRobot (movie)

Today, we have robots—or at least intelligent engines—that can write newspaper articles. Routinely, the robot produces copy about sporting events, banking information, stock markets, and a variety of other uses. The human reading the information has no idea that they have just read something produced by a machine. Humans have a bad tendency to sit around believing that there is no threat from robots because: (a) we built them and know all about them, (b) we program them to do what we want, (c) we can always pull the plug, and (d) the "laws" of robotics protect us.

Only the problem is that except for (currently) building them, none of this is true. Even the "building" aspect is already changing. We have robots that can construct other robots for a specific purpose. If the robot is a self-learning machine, then it is possible that we would not know "what it knows" if it is exposed to other sources of information. We already have robots that can do scientific research and that was in 2009.[1] Pull the plug? The DARPA designed "Energetically Autonomous Tactical Robot" or "EATR" robot forages in the fields for material to keep it functioning. Some people have claimed that the robot can also eat corpses if the need arises.[2]

Last, but not least, there are no "laws" of robotics. In fact, there are very few laws governing robotics at all. That may change in the near future with the advent of "self-driving" cars. Legislatures are already looking at ways to monitor that situation. One issue that may also appear, if we manage to develop a robot with anything approaching sentience is legal rights. Who is responsible if a robot DOES harm a human being? If a robot nurse drops the patient, should the patient sue the doctor, the hospital, the programmer, or the manufacturer? Is there liability for robotic medical procedures? What kind of rights would a robot have? Should they have any?

"Unless mankind redesigns itself by changing our DNA through altering our genetic makeup, computer-generated robots will take over our world."

Stephen Hawking

---

[1] Leslie Katz, "Robo-scientist Makes Gene Discovery—on Its Own," *CNET Magazine*, last modified April 2, 2009, http://www.cnet.com/news/robo-scientist-makes-gene-discovery-on-its-own/.

[2] John Scott Lewinski, "Military Researchers Develop Corpse-eating Robots," *Wired*, last modified July 15, 2009, http://www.wired.com/2009/07/military-researchers-develop-corpse-eating-robots/.

Shutterstock/Digital Storm

Stephen Hawking brings up an interesting point. One possible future involves combining our biological brains with robotics in such a way that we would, for example, have instant access to information. Or, we could acquire the aforementioned robotic skeleton and be far stronger. It could be argued that this is all simply a method of evolutionary behavior. Humans might be moving in the direction of melding with machines.

However, there is a dark side to this as any person who has ever seen a science fiction movie involving robots can attest. If we can build an "intelligent" device and we already know it is a device capable of thinking faster than we can, then it is only a short hop to the idea that such a device could take over. We also have science fiction to thank for the idea of a robot police force. In some respects, there are positives to the idea. It might eliminate bias (depending on the programming, of course). There is also the concept of whether we could delegate authority to a machine with firepower.

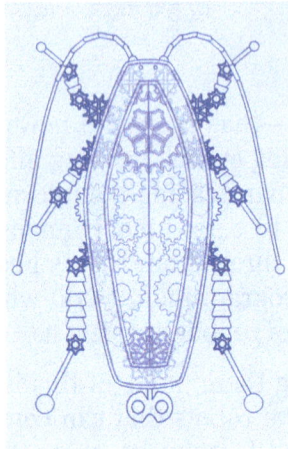

Shutterstock/John David Bigl III

Robotics began essentially with a military impetus. While one might argue that having a robot army is a truly immoral concept—we still might develop one if we knew that the enemy had one already. Today, we have drones, which have become so popular that they are available everywhere for just about any purpose. The technology behind drones was originally a military concept. The word "drone" is probably derived from the worker bees in nature. The term today is generally used to describe a vehicle of some sort that does not have a human being on board. Drones today can carry high definition cameras and can resemble almost anything. The bird that you see singing on a wire could as easily be a drone set to spy on someone. We have robotic cockroaches and we have cockroaches that you can control with a cell phone.[3]

DARPA invested money in a project called "Hybrid Insect Micro-Electro-Mechanical Systems" (HI_MEMS). Shaped like a variety of insects, these are tiny robotics that look very very real. From a distance, the differences are impossible to detect. As Stephen Colbert would say, "what could possibly go wrong with that?"

Drones currently are controlled by humans, which cause an entirely new type of problem. If you are a soldier in a war, most people are familiar with the psychological problems associated with handling a gun and killing people. With the control of a drone, you are one remove away

Shutterstock/Tyler Olson

from the action itself and this has caused some problems. While it may seem like a video game controller, you still are committing actions that cause psychological harm. Drone operators also suffer from PTSD. We have only touched the surface of this problem.

---

[3] Ian Sample, "Cockroach Robots? Not Nightmare Fantasy but Science Lab Reality," *The Guardian*, last modified March 3, 2015, http://www.theguardian.com/science/2015/mar/04/cockroach-robots-not-nightmare-fantasy-but-science-lab-reality.

DISCUSSION QUESTIONS

1. If you could have more artificial intelligence around you, what would you want?
2. Do you think a robot should have any kind of legal rights? Why or why not?
3. Are you worried that robots might mean job losses for human workers? How do you think it might affect the job market of the future?
4. How do you think a large robot population would affect our culture in general?
5. Do you think robots should be used in war? Why or why not?
6. What do you think about the ethical responsibilities are for the use of drones?
7. Are you worried about your privacy with the use of drones? Why or why not?

RESOURCES AND REFERENCES

Power, Matthew, Confessions of a Drone Warrior, GQ, 10/23/13, http://www.gq.com/news-politics/big-issues/201311/drone-uav-pilot-assassination#ixzz2j1kqr9mJ.

Danigelis, Alyssa, Cyborg Cockroach Controlled by Phone. Discovery News, 6/11/2013, http://news.discovery.com/tech/robotics/cyborg-cockroach-controlled-by-phone-130611.htm.

Dickerson, Kelly, Cyber-Roach! Mic-equipped Bugs Could Aid Disaster Rescue. Live Science, 11/7/2014, http://www.livescience.com/48676-cyborg-cockroaches-disaster-relief.html.

Chatterjee, Pratap, A Chilling New Post-traumatic Stress Disorder: Why Drone Pilots Are Quitting in Record Numbers. Salon, 3/6/2015, http://www.salon.com/2015/03/06/a_chilling_new_post_traumatic_stress_disorder_why_drone_pilots_are_quitting_in_record_numbers_partner/.

Dao, James, Drone Pilots Are Found to Get Stress Disorders Much as Those in Combat Do, NY Times, 2/22/2013, http://www.nytimes.com/2013/02/23/us/drone-pilots-found-to-get-stress-disorders-much-as-those-in-combat-do.html.

Lemonick, Michael D. Robotic Roaches Do the Trick. Time, 11/15/2007, http://content.time.com/time/health/article/0,8599,1684427,00.html.

Thompson, Mark, Unleashing the Bugs of War, Time, 4/18/2006, http://content.time.com/time/nation/article/0,8599,1732226,00.html.

Politics: Introduction

"Just as war is too important to leave it to the generals, science and technology are too important to leave in the hands of the experts."

—Sheldon Rampton

There are many people who wake up in the middle of the night with what they imagine to be a fabulous idea that will make them millions of dollars. However, the great majority of these (even the outstanding idea-types) never come to fruition. Any discussion of the intersection of science, technology, and society must include the topic of politics. Scientists would argue that we should explore space, but politicians argue whether we can afford it.

In some ways, it seems ludicrous that people with little (if any) scientific knowledge would make decisions about scientific pursuits. But it is important to note that science never occurs in a vacuum (except, perhaps, in a vacuum chamber). New developments cost money. If you think in terms of pharmacology, would it be cost effective to spend billions to design/discover a new drug to stop a disease that only a few people have? It might seem offensive to "regulate" a good deed in this fashion, but it also might seem offensive to keep from "fixing" a thousand other more common problems which would result in the greater good.

It could be argued that this is the role of politics in a society, but that is, perhaps, gilding a process that is not always altruistic. Politics is not within the scope of this book, but how politics might affect and effect scientific progress is an important aspect of the pursuit of science. One "expose" of scientific malfeasance could easily destroy the possibility of funding years of more valuable research. Imagine seeing a picture of a misshapen cloned animal (or, even more horrific, human animal). It is a short step for public response to be anti-science. Huge public outcries do tend to make politicians think twice about funding areas of development.

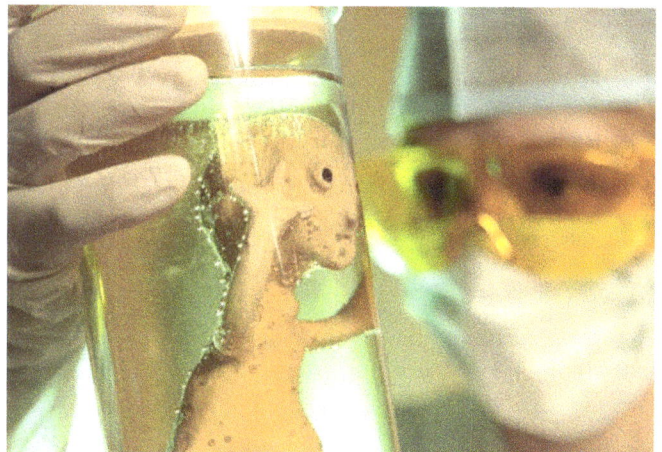

Shutterstock/Andrey Burmakin

Politics: Atoms and Chads

"The war started when people accepted the idiotic principle that peace could be maintained by arranging to defend themselves with weapons they couldn't possibly use without committing suicide."

—(Character) Julian Osborne, Movie *On the Beach*

The first nuclear bomb was dropped at Trinity White Sands (New Mexico) on July 16, 1945. As World War II wound down in 1945, a decision was made to use this gigantic bomb to effectively stop the Japanese. It was dropped on Hiroshima and, three days later, on Nagasaki in August of 1945. The "Little Boy" bomb was a uranium bomb and killed at least seventy to eighty thousand people at impact. "Fat Man" (used three days later on Nagasaki) was a plutonium bomb and killed an estimated 20 000 on impact. It is estimated that another 100 000 died afterward from health effects attributed to the bomb.[1] Sources vary on how many deaths occurred.

There is a great deal of argument (and revisionist history) over this event. It can be argued that the Japanese were about to surrender. It can be argued that dropping the bomb reduced the number of ultimate casualties if the war had continued another six months. Little information was released in America about the devastating effects of the bomb. Americans were told that the bombs had been dropped on military targets (effectively proving that either they were lousy navigators or liars—or, more charitably, that the Japanese people were "military targets"). The average person had no knowledge of the science involved.

Shutterstock/Everett Historical

While scientists were convened to "make" a bomb, there was little actual knowledge about what might happen (the concept of radiation effects was little known at the

---

[1] "Atomic Bomb-Truman Press Release-August 6, 1945," *Harry S. Truman Library & Museum*, accessed March 28, 2015, http://www.trumanlibrary.org/teacher/abomb.htm.

Shutterstock/Everett Historical

Shutterstock/vasakkohaline

United States Navy/Courtesy of the Library of Congress

time). When you watch early atomic bomb videos, you see observers standing nearby and watching. Military troops were used to test the effects of the bomb and its aftermath.[2] Huge expensive tests were designed; mock houses full of manikins with various food stuffs (some in glass, some metal, some out, and some buried) were decimated by an atomic bomb in order to see what the effects might be.[3]

While the designer scientists continued to speak out about possible aftereffects, people mostly began incorporating the idea of the atom bomb into regular life. Children of the 50s were taught (truly useless) drills like "Duck and Cover" about what to do when the bomb dropped.[4] Information about the atom bomb was even used in civil service propaganda on how to clean your house.[5] The point, here, is that it was several years AFTER the original bomb blast before we figured out what might happen with bomb radiation and its aftereffects. The politicians had a weapon they knew nothing about. They could take this amazing piece of science and affect the wished-for result: winning the war. Civilians began building bomb shelters.

Political decisions effect scientific development in many ways. Essentially, though, it has to do with a combination of "guiding" how Americans feel about science and funding what is done. We do not have a hotel on the moon yet—not because we do not have the scientific knowledge to do so—but because we have possibly better uses for our money. At one point, it was fervently believed that we "needed" the moon as a possible military base, thus we should finance such an endeavor. The original decision to "go" to the moon was based upon the idea that we could not be shown up by the Russians (coupled with the possible military aspect). Project Horizon

---

[2] "Military Participation on Operation Tumbler-Snapper (1952)|U. S. Atomic-Nuclear Bomb Tests," YouTube Video, uploaded December 2, 2013, https://www.youtube.com/watch?v=k8RSTMKB0yc.

[3] "Survival Town Atomic Bomb: Army Destroys Fabricated Town," YouTube video, uploaded November 29, 2009, https://www.youtube.com/watch?v=MG4hQQKrhT8.

[4] "Duck and Cover (1951) Bert the Turtle Civil Defense Film," YouTube Video, uploaded 11, 2009, https://www.youtube.com/watch?v=IKqXu-5jw60.

[5] "Atom Bomb Testing-The House in the Middle (1954)," YouTube Video, uploaded September 6, 2012, https://www.youtube.com/watch?v=lrYjVv9SyMQ.

was a secret military concept that was not made public for many years.[6,7]

Science and technology like the atom bomb can provide fertile ground to politicians wishing to score votes. Shortly after the end of World War II, politicians like Senator Joseph McCarthy used the paranoia engendered by the atom bomb, to scare people into believing that "the commies" were everywhere. At a speech before a small audience in Wheeling, West Virginia, he stated that "I have here in my hand a list of 205 . . . a list of names that were made known to the Secretary of State as being members of the Communist Party and who nevertheless are still working and shaping policy in the State Department. . ."[8] Saying a specific number provided a certain creditability that such a statement might not otherwise have. By the time he'd returned to Congress, he'd changed the number to 57. Even that number provoked hysteria. His committee called people to testify; they were asked if they were or ever had been a member of the Communist Party. They were also asked to name people they thought might be as well. To be called, was to have your career destroyed.

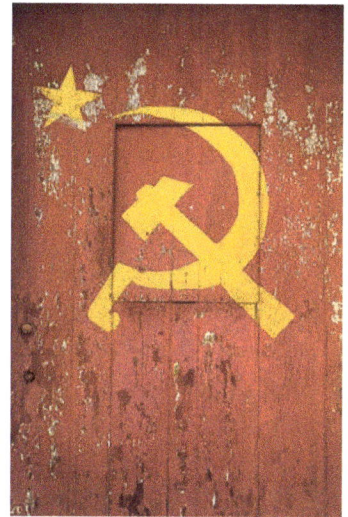

Shutterstock/Carlos Caetano

His quest included extensive television coverage, which only helped to exacerbate the situation. Many lives and careers were ruined. It ended up being a witch-hunt unparalleled in the history of the United States.

While politics unquestionably affects science and technology, the opposite is also true. The use of technology has significantly affected elections. One simple example is the idea of exit polls. If you are watching TV on election night and see that candidate X has won by a landslide (based on exit polls), you might not get out and vote even if your own polling places are still open. The infamous "hanging chad" issue in Florida is another example of technology and politics. Bush ended up winning in Florida by 537 votes (out of millions cast); the issue was related to the physical structure of the ballot. He ended up winning a majority in the Electoral College and, thus, winning the presidential election. The Electoral College itself has been the target of a possible scientific coup. There is some evidence that Internet voting (with a popular vote) could be used to determine elections instead of the somewhat archaic Electoral College concept.

When coupled with the media, films can be produced with the specified purpose of changing an election (an unsuccessful example of this would be the Michael Moore movie, Fahrenheit 9/11). The media can both lampoon and inform when it comes to political behavior. In the movie, Wag the Dog, for example, the president is caught in illicit behavior involving a young girl. The decision is made to "manufacture" a war on TV to draw attention away from the event.

Shutterstock/Lisa S.

---

6 "Project Horizon: Volume I Summary and Supporting Considerations," *United States Army*, March 20, 1959, http://www.history.army.mil/faq/horizon/Horizon_V1.pdf.

7 "Project Horizon: Volume II Technical Consideration & Plans," *United States Army*, March 20, 1959, http://www.history.army.mil/faq/horizon/Horizon_V1.pdf.

8 Joseph McCarthy, "Enemies from Within: Senator Joseph R. McCarthy's Accusations of Disloyalty," accessed March 28, 2015, http://historymatters.gmu.edu/d/6456.

DISCUSSION QUESTIONS

1. Do you (or your family) have any sort of disaster plan in case of a national emergency? If so, what?
2. Do you think that you need such a plan? Why or why not?
3. Can you imagine such a scenario now? In what circumstances?
4. Do you think that profiling terrorists is similar to the behavior of Senator McCarthy?
5. What do you think has changed about the political culture that makes us more susceptible now?

RESOURCES AND REFERENCE ARTICLES

Cosgrove, Ben. The Haunted Desert: Aftermath of a Nevada A-Bomb Test, Time, 5/30/12, http://time.com/3675016/nevada-a-bomb-test/.

Hiroshima Atomic Bomb (1045) https://www.youtube.com/watch?v=t19kvUiHvAE.

Survival Town Atomic Bomb Test. Army Destroys Fabricated Town https://www.youtube.com/watch?v=MG4hQQKrhT8.

Operation Tumbler-Snapper (1952) https://www.youtube.com/watch?v=k8RSTMKB0yc.

'60s minute: The Space Race, CNN, 8/25/2014 http://www.cnn.com/2014/07/24/us/1960s-moon-military-base/.

Politics: Censorship

"Ethics change with technology."

—Larry Niven, N-Space

We know more today about the people running for office. A hundred years ago, we had little knowledge of the social or sexual behaviors of our politicians. Obviously some people knew, but knowledge like that was easily suppressed (or poorly spread). The people who voted primarily voted along certain basic lines. Would this candidate represent my interests? Would this candidate do a good job? Was he a fine fellow? Scandals were poorly reported (if at all) because it was deemed "unseemly" for the most part. As we moved into the twentieth century, information was suppressed for a variety of reasons.

During World War II, for example, it was "hidden" that Franklin Delano Roosevelt was essentially motor-impaired from a bout with infantile paralysis (polio). He had been diagnosed in 1921 (he was 39 years old) but it was a political liability at the time.[1] After he became president, the general idea here was that it would make him appear "weak" and would thus discredit our power standings among the nations of the world. There were nearly no pictures of him showing any degree of impairment. The fact that he routinely used a wheelchair and heavy leg braces was simply not shown. He made routine "whistle-stop" train trips around the country to promote his agenda. The train would move from small town to small town and he would appear at the back of the caboose to give a speech. He had to be all but "tied" to the rail to stand upright, but you were not allowed to make any public mention or photograph of that fact. In addition, it was routine to have the secret service holding machine guns stand in front in a semicircle to protect him. Again, you weren't allowed to take pictures of that (Eugene Cowert, pers. comm. to author Potter, 1970's).

Shutterstock/dustin77a

Today we know that he had an affair (he was, in fact, with her when he died),[2] but at the time, that information was well hidden. It detracted from his "heroic" image. Today, that simply no longer occurs. Once disability rights became a popular mantra,

---

[1] Amy Berish, "FDR and Polio," *Franklin D. Roosevelt Presidential Library and Museum*, accessed 28 March, 2015, http://www.fdrlibrary.marist.edu/aboutfdr/polio.html.

[2] Charles McGrath, "No End of the Affair," *The New York Times*, April 20, 2008, http://www.nytimes.com/2008/04/20/weekinreview/20mcgrath.html?pagewanted=print&_r=0.

Shutterstock/Simfalex

Shutterstock/nito

it became "okay" to have a sculpture of him in a wheelchair—or at least in a wheelchair obscured by a cloak.

While we might try to hide scurrilous information, this is the age of the Internet where anyone can post anything. The sad fact, though, is that so much information is posted that it becomes a brief "wonder" that subsequently disappears with the next scandal that comes along.

The "Mister Smith Goes to Washington" model of a good man that goes off to help us in the halls of Washington no longer applies. We have little or no respect for our elected officials (probably for good reason). We elect them based upon whether their worldview and rhetoric match our own (causing even more political polarization).

In 2015, we have an elected official asking a doctor if you could swallow the "camera" that is used for a colonoscopy to look and see if there is a baby "in there." People who support this congressperson do not seem to care that he has little or no knowledge of basic human anatomy. They care that he has the same point of view on the issue as they do. The issue of how they are funded also seems to be of little concern to many people.

Science and technology are no longer the answer. They aren't even the question. Today, we want to follow people who agree with us. It is a very interesting concept to consider. If scientific "fact" cannot convince you of a specific aspect of an issue, then what exactly could? The answer, sadly

Shutterstock/Jiri Miklo

---

3 "Followup to the Camera-Comment Controversy," *Lowering the Bar*, accessed March 28, 2015, http://www.loweringthebar.net/2015/02/followup-to-the-camera-comment.html.

to say, is probably nothing. People tend to decide things based more upon their common experience than on any variety of scientific fact. In some cases, that has some validity. Science has certainly backtracked (back AND forth) on various issues. Coffee will kill you. Coffee will make you live longer. Give up butter and eat margarine. Wait, margarine will kill you. If you are completely inundated with "facts" from one scientist or another, it is almost to be expected that sooner or later what you believed to be true before will resurface. If grandpa smoked a pack of unfiltered cigarettes every day and died at 95, you might ignore the warnings on cigarette packages.

Shutterstock/Brian A. Jackson

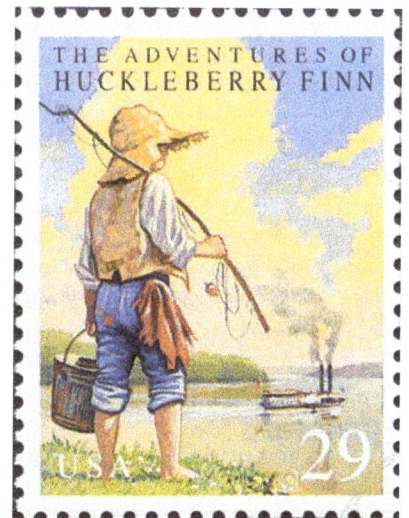

Thus we can elect someone who clearly does not know anything about science as long as he or she agrees with us. The Internet has only made this easier. But should we be "hiding" any information? If you ask that straightforward question, most people will answer, "of course not." But the situation is a bit more complicated. Do you want to see images of brutality or violence or sex? Perhaps you do and perhaps you think you should be able to see it. However, most societies will start to "draw lines" when it comes to the availability of such with children, for example.

Censorship is difficult. In his movie, Fahrenheit 9/11, Michael Moore wanted to take pictures of the coffins of military soldiers coming back from war (he was denied; of course he was also most likely illegally on the military base at the time). You can have a knee jerk reaction to this—"no, that's disrespectful to the dead." First off, they are anonymous (you don't know who is inside). And second, does seeing a huge set of them make you think twice about war? Not an easy conundrum.

In the movie (and book) Fahrenheit 411, a dytopian future is portrayed. In this future, all books are burned. It is considered better for you not to know things. You are happier not knowing. You conform better. Absurd thought the concept is, book banning is an interesting problem as well. Do we or should we ever prevent a child from seeing a certain book?

In 2011, Alan Gribben, a professor at Auburn University in Alabama edited the Mark Twain's books, The Adventures of Tom Sawyer and The Adventures of Huckleberry Finn to remove a racially derogatory "N" word.[4] The professor, who is Caucasian, caused a firestorm of criticism over the move. On the face of it, it seems horrific—the idea that you could change a classic (which was, relatively speaking, fixed in an earlier historic time when such language was acceptable). In this case, however, his rationale is interesting. The book, apparently, could not be stocked in school libraries because of the use of that word. This is an interesting dilemma. Perhaps we should look at the underlying original idea but it is a problem that has merit on both sides of the issue.

Shutterstock/catwalker

_____

4 Alec Harvey, "Auburn-Montgomery professor Alan Gribben Not Shocked his Editing of Twain Classics Drawing Fire," AL.com (blog), January 5, 2011 (10:40 a.m.), http://blog.al.com/scenesource/2011/01/auburn-montgomery_professor_al.html.

DISCUSSION QUESTIONS

1. What do you think about your politicians? Do you think there are still good honest people worthy of respect? Why or why not?

2. Do you agree or disagree with the censorship of a book like Tom Sawyer or Huckleberry Finn?

3. Do you think older works should be "censored" to fit modern sensibilities about race and culture? Why or why not?

4. Is it better for a child to have access to more information or for that information to be censored for appropriateness based on age?

5. When do you think we should have censorship (if at all)?

RESOURCES AND REFERENCES

Alan Gribben videos, podcasts on Twain's Tom Sawyer and Huck Finn http://www.newsouthbooks.com/pages/2011/07/15/alan-gribben-videos-podcasts-on-twains-tom-sawyer-and-huck-finn/.

FDR—American Experience http://www.pbs.org/wgbh/americanexperience/films/fdr/player/.

Mini-Bio Franklin Delano Roosevelt https://www.youtube.com/watch?v=qK42SUseTwM.

Medicine and Health: Introduction

"The aim of medicine is to prevent disease and prolong life, the ideal of medicine is to eliminate the need of a physician."

—William James Mayo[1]

The history of medicine goes back many years. It probably began with the accidental use of plants for medicinal purposes. The Caves of Lascaux show examples of this use dating over 15,000 years ago (on a side note, The Caves of Lascaux also include alleged pictures of unicorns). While today, we have medical tests and expensive technology, the basic tenets of medicine have remained essentially the same. A person gets ill and seeks aid. However, beginning in the twentieth century, the idea of a "cure" for any and all diseases was superseded by the idea of prevention. The concept of prevention, of course, had existed for thousands of years, but, generally speaking, one did not go to the doctor to prevent disease. You went to the doctor to be made well.

A fortuitous accident in 1928 led to the discovery of penicillin by Alexander Fleming. It was a total accident, due primarily to the disarray in his lab, that he discovered penicillin could kill off staph bacteria. He did not have the money or the methodologies of manufacturing it in large quantities and it was about 10 years later before it came to be commercially produced and used.[2] This ushered in a period where people believed that the future of medicine was in the use of these new miracle drugs. Unfortunately, especially with the discoveries concerning viral diseases, a "cure" was not always possible.

The major epidemics of the twentieth century brought home the idea that modern medicine did not have all of the answers. The "swine" flu epidemic of 1918 and 1919 resulted in approximately 50 million deaths worldwide.[3] It was a strange new type of disease in that it targeted and killed off healthy people aged 20–50 (not the usual old people and babies of prior flu epidemics). In 1952, at the high point of

---

[1] William James Mayo, "The Aims and Ideals of the American Medical Association," *Proceedings of the 66th Annual Meeting of the National Education Association of the United States* (1928): 163.

[2] "People and Discoveries: Fleming Discovers Penicillin 1928–1945," *PBS.org*, last modified 1998, http://www.pbs.org/wgbh/aso/databank/entries/dm28pe.html.

[3] U.S. Department of Health & Human Services, "Pandemic Flu History,"*Flu.gov*, accessed April 1, 2015, http://www.flu.gov/pandemic/history/.

the infantile paralysis (polio) epidemic, there were 58,000 cases and 3,145 deaths.[4] These were diseases that had no cure and there was no cure in sight. We began the inexorable march toward acquiring vaccines to prevent diseases. We had vaccines before that, but at this point, we decided that it was the most effect method of treatment. It is cheaper to prevent than to cure.

Medicine began to shift away from the god-like physician who cured.

---

[4] "The 10 Worst Disease Outbreaks & Epidemics in U.S. History," *Healthline.com*, last modified January 20, 2013, http://www.healthline.com/health/worst-disease-outbreaks-history#SpanishFlu5.

Medicine and Health: Vaccine Science

FLATOW: Chantal, is there anything that Dr. Offit could tell you to change your mind?

CHANTAL: Absolutely not.

FLATOW: So even if his studies are correct, and you admit or go back and research what he's saying to be true?

CHANTAL: I do not put any faith into anything that my government tells me.[1]

CHAPTER 8b

According to vaccines.gov, "A vaccine is a product that produces immunity from a disease and can be administered through needle injections, by mouth, or by aerosol."[2] The idea behind vaccines is based upon a very old idea of "inoculating" someone with a tiny part of the disease (today, usually a killed part) in order for your body to create its immunity to the disease. The first active use was with smallpox and fortunately, it was discovered that a lesser variety (cowpox) might produce the necessary antibodies.

Virtually nowhere in medical science is there more controversy than there is about vaccines. Vaccine safety experts claim test after test, experiment after experiment, and case after case that vaccines are, for the majority of people, very safe methods of curtailing disease. They use the horror stories of the past to establish sympathy, but they fervently claim to base their decisions on scientific research. On the other side, the anti-vaccination people are vocal in their opposition to vaccines. The anti-vaccine people produce extremely technical and effective videos to make their cause. The videos tend to be full of horrific cases of sick, injured or dead children to prove that vaccines are dangerous. Finally, there are those who are not anti-vaccine, but who do take issue with CDC recommendations on the number of vaccines and the vaccine schedule. These parents would prefer an alternative vaccination schedule built largely of single dose vaccines. Both the CDC and the American Academy of Pediatrics are opposed to this idea.[3]

If we consider our prior chapters concerning logical fallacies, add to that the idea that beautifully constructed videos help you to buy-in, and have your Internet give you the answers you want, you can see how many people truly believe that vaccines

---

[1] "Paul Offit on the Anti-Vaccine Movement," *NPR.org*, last modified January 7, 2011, http://www.npr.org/2011/01/07/132740175/paul-offit-on-the-anti-vaccine-movement.

[2] U.S. Department of Health and Human Services, "Glossary of Vaccines and Immunizations Terms," Vaccines.gov, accessed on April 1, 2015, http://www.vaccines.gov/more_info/glossary/index.html#v.

[3] John Snyder, "Cashing in On Fear: The Dangers of Dr. Sears," *Science-Based Medicine*, 7/20/09, https://www.sciencebasedmedicine.org/cashing-in-on-fear-the-danger-of-dr-sears/

are dangerous. They can even find experts to agree with them. The pro-vaccination people like to attack these experts as not really all that expert in the appropriate field (another logical fallacy).

Feeding into this controversy also is the problem that virtually no one of child-bearing age has ever seen or experienced the diseases of childhood that vaccines prevent. If you have never had measles or whooping cough, they can be listed as "typical" childhood diseases. The anti-vaccine argument is that having the disease produces a natural immunity, which is safer than a "chemically induced" immunity provided by a vaccine.

There are several problems with that concept. In a variety of scientific studies, there is little or no evidence that the "chemistry" of vaccines is dangerous. You can cite the ingredients (which can be compelling in their naming alone), but the actual amounts are extremely minimal (as the vaccine safety people would say, you get more mercury in fish). The second problem is that if you actually expose someone to a childhood disease like measles, you are running the risk of some extremely severe complications like encephalitis. Are those cases common? No. Can they be fatal? Yes. Even if the odds are very small, it is a risk most people who have seen the disease believe to be important.

One other problem exists with the idea of correlation and causation. If you have an immunization today and two weeks later you have a stroke, it is impossible to prove that the vaccine caused the stroke (what are the odds you would have had a stroke in the first place?). There are, in fact, documented cases of vaccine injury, but they generally occur because of underlying conditions that were unknown at the time of the vaccination.

The idea that vaccines cause autism has been scientifically studied to such a degree, that there is, emphatically, no identified causation between autism and vaccines. You can point to the rise in people diagnosed with autism, but this could be due to a large variety of factors unrelated to vaccines including previously unidentified and seemingly unrelated illnesses, conditions, or diseases. The original "study" used to establish this belief was conducted by Andrew Wakefield. After publishing the results of a study, the article was retracted by the Lancet (where it was published) due to erroneous information. In addition, Dr. Wakefield lost his medical license because of it.[4] Unfortunately, Dr. Wakefield continues to lecture on the anti-vaccine circuit.

The original study used to "prove" this claim was based on 12 children. The extremely small sample was questionable but there were also a number of discrepancies in the study. Some of the children already had developmental issues; some were paid. Ten of the thirteen authors retracted their story. There were multiple discrepancies found.[5] Unfortunately, the fact that the story was not true has had little or no effect on the anti-vaccination movement. As is sometimes the case in the court of public opinion, they believe that the retraction was some sort of cover-up.

The U.S. government established the Vaccine Injury Compensation Program in 1988 to protect American citizens from possible injuries from vaccines. Given that not everyone reports injuries, the list of side effects is not comprehensive but

---

[4] Clyde Haberman, "A Discredited Vaccine Study's Continuing Impact on Public Health," *NYTimes.com*, last modified February 1, 2015, http://www.nytimes.com/2015/02/02/us/a-discredited-vaccine-studys-continuing-impact-on-public-health.html?_r=2.

[5] Brian Deer, "Secrets of the MMR Scare: How the Case Against the MMR Vaccine Was Fixed," *BMJ* 342 (2011): c5347, accessed April 1, 2015, https://owl.english.purdue.edu/owl/resource/717/04/.

effectively serves the same purpose due to the extensive reporting. You can look online and see how many people had problems with each vaccine and what those problems were.[6] The trust fund was established to compensate people in the case of vaccine injury; however, it should be pointed out that it also protected pharmacological companies from large payouts. Vaccines are not always cost-effective and companies might refuse to produce them if they are no longer profitable.[7]

Unfortunately, it is very difficult to verify specifically if a vaccine causes harm. Vaccine production is another problem. Historically, there has been at least one major vaccine production incident at the start of the polio vaccine development. Known as the "Cutter Incident," in 1955, a company in California produced a vaccine that caused harm. "In the end, at least 220,000 people were infected with live polio virus contained in Cutter's vaccine; 70,000 developed muscle weakness, 164 were severely paralyzed, and 10 were killed."[8] Today, vaccine production is monitored very closely. In addition, there was a negative impact in that an oral polio vaccine was touted as more effective (and safer) when, in fact, it was not.[9]

The idea behind vaccination science is that if you can get a large number of people vaccinated, then it will protect those in a given population who cannot get vaccinated. This concept, called "herd immunity," is the basis for modern vaccination laws. If a large area/group decide not to get vaccinations, then it is possible that the vaccine-preventable diseases might spread and cause harm. There have been several such incidents in recent years. One case involved Disneyland in California where a measles epidemic occurred. Another case involved a cluster of pertussis (whooping cough) cases (also in California).

Vaccines bring us ethical issues as well. Can we, in good faith, "require" people to immunize their children for the better good? Today, 50 states allow for medical exemptions, 48 allow for religious exemptions, and 20 allow "philosophical" exemptions.[10] Legislation is also pending in several states to change or eliminate some exemptions.

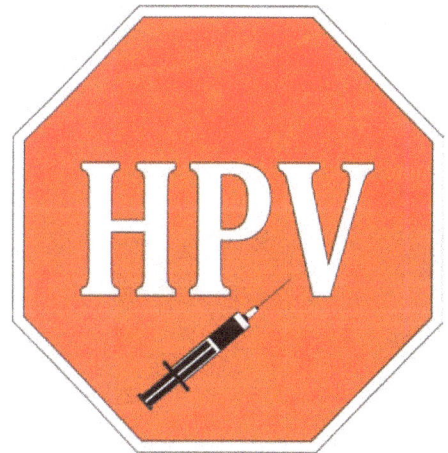

Shutterstock/Sangoiri

This is a difficult area. Most parents want what is best for their children, but many parents do not want to be forced to a certain behavior. It is one of the better cases where modern science and scientific proof has not always overcome personal doubt.

In 2006, the FDA approved a vaccine against the Human Papilloma Virus (HPV). On paper, this was excellent news. HPV is one of the most common sexually transmitted infections in

---

[6] U.S. Department of Health and Human Services, Health Resources and Services Administration, "National Vaccine Injury Compensation Program," *HRSA.gov*, accessed April 1, 2015, http://www.hrsa.gov/vaccinecompensation/index.html.

[7] The College of Physicians of Philadelphia, "The History of Vaccines," *Historyofvaccines*.org, last modified July 31, 2014, http://www.historyofvaccines.org/content/articles/different-types-vaccines

[8] Paul A. Offit, *The Cutter Incident: How America's First Polio Vaccine Led to the Growing Vaccine Crisis* (New Haven, CT: Yale University Press, 2005), p. 89.

[9] Teri Shors, Understanding Viruses (Jones & Bartlett, 2011), p. 333.

[10] The College of Physicians of Philadelphia, "The History of Vaccines," *Historyofvaccines*. org, last modified July 31, 2014, http://www.historyofvaccines.org/content/articles/different-types-vaccines.

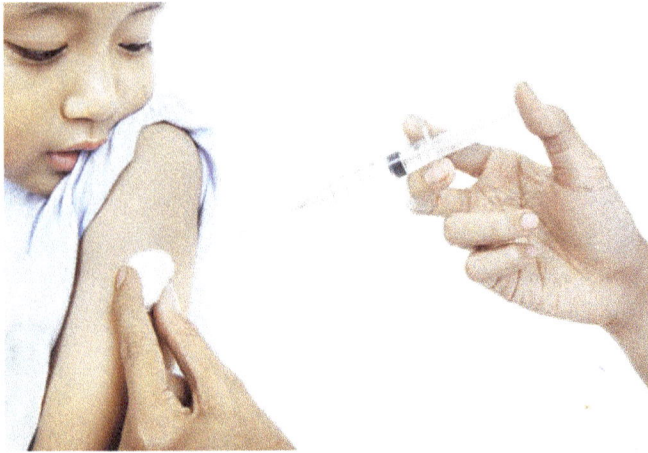

Shutterstock/Creativa Images

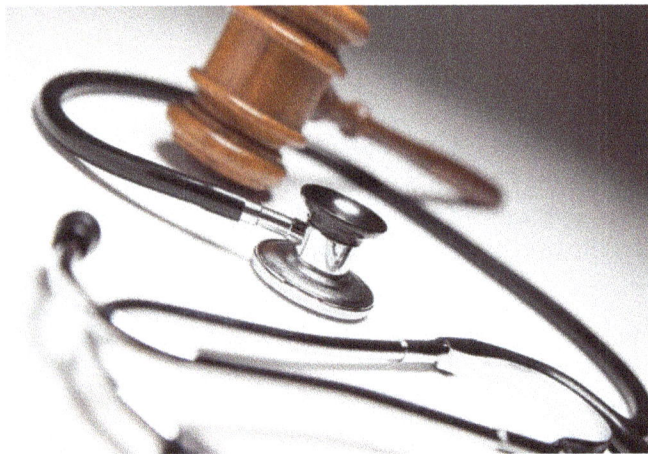

Shutterstock/Andy Dean Photography

Shutterstock/Marcos Mesa Sam Wordley

adults. There is evidence that it can lead to several different cancers (primarily cervical cancer). The development of a vaccine to prevent any kind of cancer is an amazing step forward.

However, this vaccine, to be effective, has to be given to children before the age of sexual activity (it is no longer effective once you have been exposed). Many parents balk at the idea of giving a vaccine against a sexually transmitted infection to a child. They even put forth the idea that giving such a vaccine will make the child more likely to be sexually promiscuous. Of course, the idea that any parent would actually sit down and discuss with any child the exact meaning and point of every vaccine is a bit specious. How many people have heard someone say, "Listen, dear, if you will only agree to getting this shot, you will never get diphtheria/pertussis/rubella?"

It is, however, yet another example where medical ethics and vaccines collide. Few parents want to hurt their children and most want to make the best-informed decisions that they can. With the abundance of available material online coupled with the persuasive flair of the Internet can cause people to question scientific information.

There is also a problem with the way vaccines are marketed. Each year, in the United States, we have a "new" flu vaccine based upon the anticipated strains of flu to come that year. Some years, they are very effective, some not so much. Because of that, way too many people refuse to get the vaccine (and way too many people believe, erroneously that it will give them the flu). Public health is worried that there might be another flu epidemic like the "swine" flu epidemic of 1918–1919, but people today believe that either it will never happen or modern medicine will keep them alive. Both are dangerous concepts.

Today, we have moved beyond even the preventive idea of vaccines. We now have recommended practices that encourage a diagnosis of "pre-hypertension" or "pre-diabetes." It has become the norm to visit a doctor for a routine physical in order to see the "pre" possibilities of disease. Part of the reason for this is the epidemics we could not cure, but in addition to that, there was also a cost factor. It's cheaper to get someone to live healthier than it is to cure disease.

DISCUSSION QUESTIONS

1. How do you feel about vaccines in general? Do you think they save lives? Why or why not?

2. Do you and/or the members of your family get a flu shot? Have you ever had the flu? Do you think we can effectively prevent the flu?

3. Do you believe that the government or pharmacology companies are hiding information and deliberately trying to hurt you? If so, why?

4. If you search for information about vaccines, are you swayed by the abundance of anti-vaccination propaganda? Why or why not?

5. If vaccines are safe, what do you think is the most effective way of disseminating information?

RESOURCES AND REFERENCES

Itzkowitz, Harold. Film Review: When there was no vaccine. The National Memo, 2/6/2015, www.nationalmemo.com/film-review-vaccine/.

Wakefield, AJ, et al., RETRACTED: Ileal-lymphoid-nodular hyperplasia, non-specific colitis, and pervasive developmental disorder in children. The Lancet, Vol 351, No. 9103, pp. 637–641, 2/28/1998, http://www.thelancet.com/journals/lancet/article/PIIS0140-6736(97)11096-0/abstract.

Vaccine Injury Table, http://www.hrsa.gov/vaccinecompensation/vaccinetable.html.

Data and Statistics, http://www.hrsa.gov/vaccinecompensation/data.html.

The History of Vaccines, The College of Physicians of Philadelphia, http://www.historyofvaccines.org/content/articles/different-types-vaccines.

Offit, Paul A., The Cutter Incident: How America's First Polio Vaccine Led to the Growing Vaccine Crisis. Yale University Press, 9/28, 2007, CDC—HPV, http://www.cdc.gov/hpv/vaccine.html.

Vaccines—Calling the Shots

PBS Nova, 9/12/2014 http://www.pbs.org/wgbh/nova/body/vaccines-calling-shots.html.

Hoax Video of Girl Walking Backwards as a result of the HPV Vaccine https://www.youtube.com/watch?v=5ztiAN9k584.

Polio Vaccine Trials Begin http://www.history.com/this-day-in-history/polio-vaccine-trials-begin.

The Future of Immunization http://www.historyofvaccines.org/content/articles/future-immunization.

Medicine and Health: Medical Ethics

"Only one rule in medical ethics need concern you—that action on your part which best conserves the interests of your patient."

—Martin Fischer

In 2005, the category 5 hurricane named Katrina struck New Orleans. Between the hurricane and the storm surge, nearly 2000 people died. Natural disasters are a good starting point for a discussion of medical ethics.

Shutterstock/Guido Amrein, Switzerland

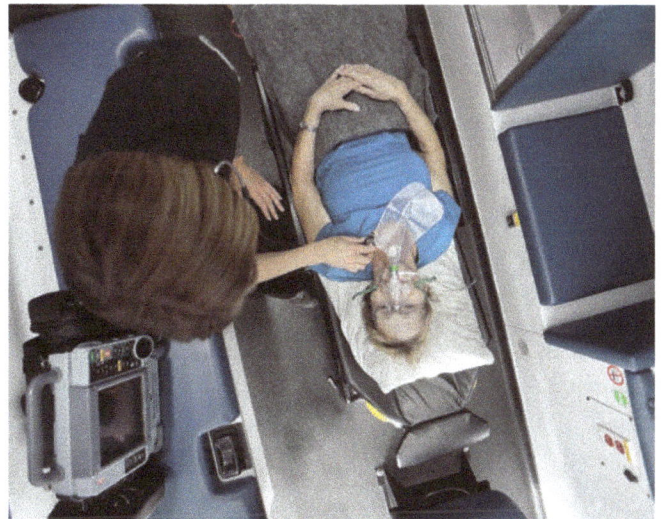

Memorial Hospital, at the point of Katrina's arrival, had approximately 200 patients and an additional 1400 people taking shelter from the storm. The power went out early in the morning of August 29, 2005. The generators kicked in, but they were never designed to run anything other than temporary needs. The air conditioning cut off. It was over a hundred degrees and there was limited lighting. The next day, people in the hospital were horrified to realize that the water was rising. Rising water might mean that the electrical system would shut down entirely.

Memos went out in an attempt to begin an orderly evacuation of patients. At that time, Memorial had a "wing" devoted to critically ill long-term care patients. This was an independent unit within the hospital.

People making decisions had to determine who would be evacuated first. The concept of "reverse" triage took effect. Reverse triage is based upon a military concept; it is the idea that you should evacuate the people most likely to live. It is the reverse of normal hospital triage where medical personnel see the most ill patients first.

Some of these long-term care patients were on ventilators. Some of the patients in the hospital were marked "DNR" (which stands for

Shutterstock/Tyler Olson

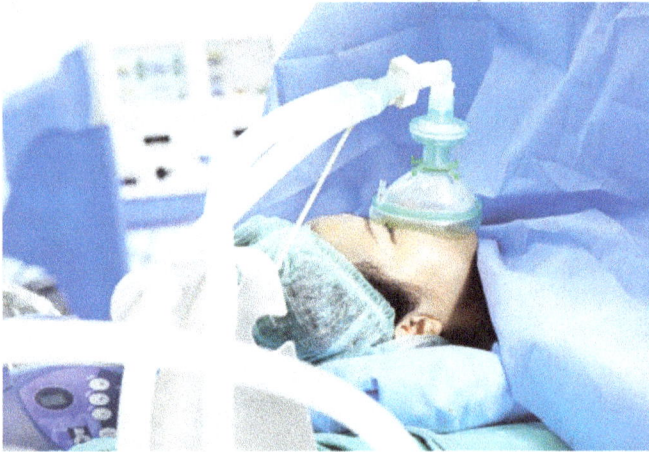

Shutterstock/Apples Eyes Studio

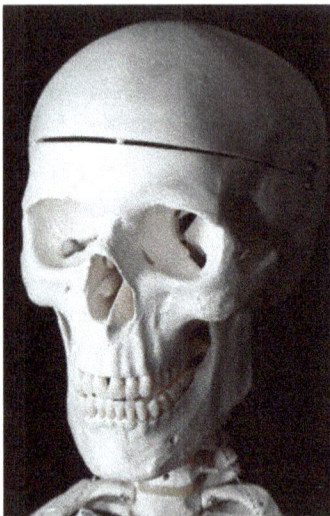

Shutterstock/Pixel 4 Images

Shutterstock/kamnuan

"do not resuscitate"). Once the power failed, any care would have to be manually provided. Since the elevators were not functioning, some of these patients had to by physically carried down multiple flights of stairs in the dark. Several died while awaiting evacuation. Through an enormous collision of poor communication skills, these people were somehow omitted from the evacuation plans.

In July of 2006, Louisiana investigators arrested Dr. Ann Pou and several nurses in connection with the deaths of these patients. The charge was that these caregivers had chosen to hasten the death of certain patients with the administration of morphine. They were not indicted by the Grand Jury. While it is impossible to completely verify what happened, there were at least 17 patients with morphine in their system during autopsy.[1]

That is the crux of the problem with medical ethics. If you can only save a certain number of people, is it "fair" or right to end the lives of people who would not survive under the guise of personal dignity? Medical ethics is a set of moral values and belief structures that should guide healthcare. Some cases, like Katrina, give clear-cut issues. We might agree with one side in particular, but all sides have merit.

Historically, the medical establishment has made several missteps when it comes to deciding on medical research. It is sometimes easy to be misled by the concept of the greater good. If I do X research on a limited number of people, then my research might save a much larger number of people in the future. Once upon a time, it was believed that a lobotomy could be a miracle cure for the severely mentally ill. A lobotomy involves cutting into the fontal lobe of the brain and making a few strategic cuts to nerves. Did it make the patient "feel" better? It definitely made them calmer, but it was never a cure. Like the concept of electroshock therapy, it might make the patient behave better (and possibly even feel better), but it still caused damage.

If you believe, for example, that certain races of people are less important, than others, it is not a huge step to justify doing medical experimentation on them. Eugenics, a popular "science" of the nineteenth and twentieth centuries promoted the idea of breeding "better" characteristics to provide stronger/better humans. It took the experimentation of the Nazis to put the final death knell to this concept.

---

[1] Sheri Fink, "The Deadly Choices at Memorial," *NYTimes.com*, last modified August 25, 2009, http://www.nytimes.com/2009/08/30/magazine/30doctors.html?pagewanted=all&_r=0.

In 1932, the Public Health Service began studying syphilis in Tuskegee, Alabama. The "idea" was to figure out how to treat and provide help to future sufferers of syphilis by effectively measuring the untreated progress of the disease. The study included 600 black males; 399 of them had syphilis and 201 did not.[2]

There were two major ethical problems with the study. First, they were not even told what the study was about; there was no "informed consent" to agree to the study. And second, the participants did not receive treatment during the FORTY YEAR study. By 1947, these men could have been treated with penicillin. They were not. The study finally ended in 1972 due to a public outcry.[3]

While conspiracy theorists are convinced that the government is "out to get you," they may have some grounds for this theory when considering activities such as the secret CIA Mkultra project. Designed to study behavior modification, subjects were given hallucinatory drugs and monitored.

During Prohibition, the U. S. government decided to make denatured alcohol poisonous in an effort to make people stop drinking. This "scare tactic" ended up killing thousands of people.[4]

Another important question has to do with where we draw the line in medical care. At what point can a patient say, "no more" and have it accepted? How does that conflict with the doctor's mission to heal? Dax Cowart was a young air force reservist. In 1973, he accompanied his father to visit a piece of property. Due to a gas leak, there was an explosion. Both men were severely burned; the senior Cowart died on the way to the hospital. Dax told everyone who would listen that he did not want to live and he did not want anything done to keep him alive. Psychological testing was done; it established that he was "in his right mind" and could make that decision. His mother and his doctors refused his request. His treatment involved being dipped in a chlorine bath, his skin had to be repeatedly "sanded" and

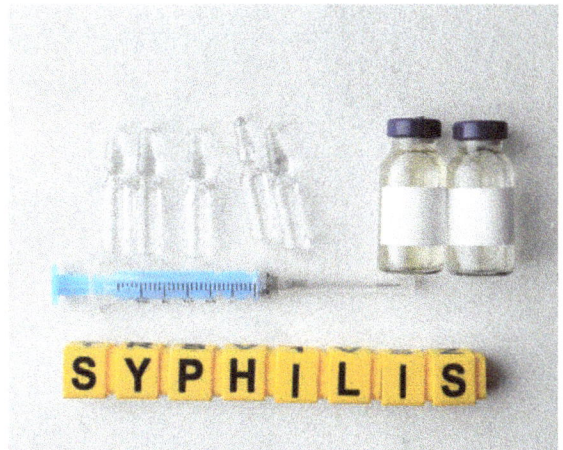

Shutterstock/Africa Studio

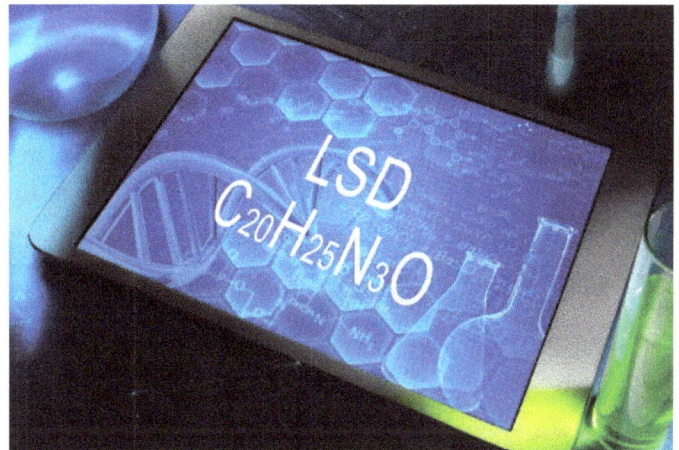

Shutterstock/designer491

Shutterstock/Everett Historical

---

[2] Centers for Disease Control and Prevention, "U.S. Public Health Service Syphilis Study at Tuskegee: The Tuskegee Timeline," *CDC.gov*, last modified September 24, 2013, http://www.cdc.gov/tuskegee/timeline.htm

[3] Ibid.

[4] Deborah Blum, "The Chemist's War: The Little-told Story of How the U.S. Government Poisoned Alcohol During Prohibition," *Slate*, February 19, 2010, http://www.slate.com/articles/health_and_science/medical_examiner/2010/02/the_chemists_war.html.

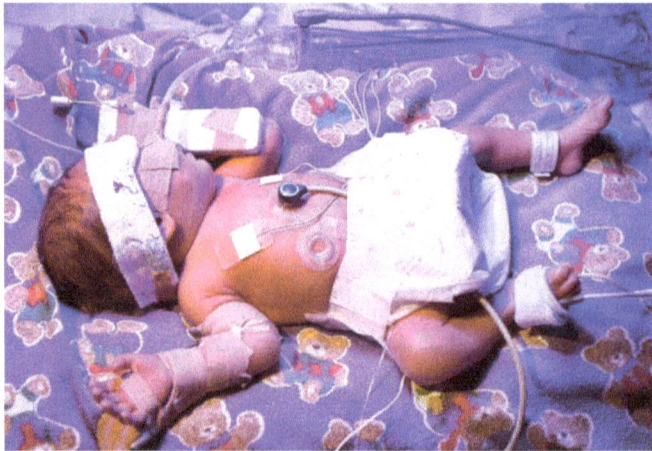

Shutterstock/Steve Lovegrove

the pain was immense. He lived. He was blind and had nonfunctioning hands. He went on to acquire a law degree and represents patients in similar situations.[5]

Medical ethics is an important crossroads for any discussion of the impact of science and technology on human beings. You can see similar problems if you look at the ideas surrounding the saving of extremely premature infants. The statistics show the grim outcomes and financial costs involved. Just because we CAN save a baby weighing less than a pound, should we? It only takes one "success" story before people begin believing that it is always a good idea. Some argue that life is so precious that we must always preserve it; others take what they consider to be a more humane approach—life with horrific suffering is not viable life. It is a scientific conundrum with all the aspects of humanity and technology tied irrevocably together.

On the other end of the spectrum, the world is aging at a very rapid rate. Today, modern medicine keeps us alive longer. If you believe Ray Kurtzweil, the possibility of living forever is within our grasp. As with preemies, there comes a point in which we have to consider whether it is humanely or financially viable to keep people alive. Ever-increasing numbers of the elderly means major change in our cultural structure. We do not have the resources at this time to meet the needs of our current elderly.

Shutterstock/Ollyy

Medical ethics gives us the last puzzle piece. It is not always a question of what we CAN do. Sometimes, it is a question of what we should ethically and morally do. Technology can provide a solution; it is not always the best solution.

---

5 University of Washington, Ethics in Medicine, last modified September 3, 2014, https://depts.washington.edu/bioethx/tools/ceintro.html.

DISCUSSION QUESTIONS

1. What form of triage do you think should happen in an emergency? Why?

2. Is it "fair" to ask a hospital worker to abandon their own family to help in an emergency?

3. If it was your relative in the hospital in terrible pain with no possibility of treatment, what choice would you make?

4. Do you think the patient is always right when it comes to his or her own body?

5. Do you want to live forever?

RESOURCES AND REFERENCES

Knabb, Richard D; Rhome, Jamie R; Brown, Daniel P; National Hurricane Center (December 20, 2005). Hurricane Katrina: August 23–30, 2005 (PDF) (Tropical Cyclone Report). United States National Oceanic and Atmospheric Administration's National Weather Service, http://www.nhc.noaa.gov/pdf/TCR-AL122005_Katrina.pdf.

Fink, Shirley. Five days at Memorial; Life and Death in a Storm-ravaged Hospital. Crown Publishers, NY, 2013

Fink, Shirley. The Deadly Choices at Memorial.ProPublica, 8/27/2009, http://www.propublica.org/article/the-deadly-choices-at-memorial-826.

Pernick, Martin S. The Black Stork; Eugenics and the Death of "Defective" Babies in American Medicine and Motion Pictures Since 1915. Oxford University Press, NY 1996.

Principles of Medical Ethics, American Medical Association, revised June 2001, http://www.ama-assn.org/ama/pub/physician-resources/medical-ethics/code-medical-ethics/principles-medical-ethics.page.

U.S. Public Health Service Syphilis Study at Tuskegee, The Centers for Disease Control, http://www.cdc.gov/tuskegee/timeline.htm.

Brandt, Allan M., Racism and Research: The Case of the Tuskegee Syphilis Study, http://www.med.navy.mil/bumed/Documents/Healthcare%20Ethics/Racism-And-Research.pdf.

The troubled history of the foreskin, Mosaic, 2/24/2015, http://mosaicscience.com/story/troubled-history-foreskin.

Project Mkultra, The CIA's Program of Research in Behavioral Modification, Joint Hearing before the Select Committee on Intelligence and the Subcommittee of Health and Scientific Research of the Committee on Human Resources, the U.S. Sentate, 8/3/77, http://www.nytimes.com/packages/pdf/national/13inmate_ProjectMKULTRA.pdf.

Blum, Deborah, The Chemist's War: The Little Told Story of How the U. S. Government Poisoned Alcohol During Prohibition. Slate, 2/19/2010, http://www.slate.com/articles/health_and_science/medical_examiner/2010/02/the_chemists_war.html.

van Bogaert, Knapp et al, Assistance in Dying: Dax's Case and Other Reflections on the Issue. Ethics CPD Article 5. Vol 52 #6, http://www.che.org/members/ethics/docs/2666/Dax%20Cowart.pdf.

Never Let Go. Tampa Bay Times article series on prematurity, http://www.tampabay.com/specials/2012/reports/juniper/.

Reporting Alleged Euthanasia: "The Deadly Choices at Memorial" https://www.youtube.com/watch?v=DmLY6K7gAUk.

Watch This 1950s Woman Get Dosed with LSD https://www.youtube.com/watch?v=iGf2loLAwVE

CIA Medical Experiments: Treating Psychosis—MKULTRA Mind Control Documentary Film (1955) https://www.youtube.com/watch?v=NMbvZtBTRb8.

Dax Cowart 2002 1 https://www.youtube.com/watch?v=lSsu6HkguV8.

Dax Cowart https://vimeo.com/64585949.

Preemies in the NICU: A Visual Tour http://www.babycenter.com/2_preemies-in-the-nicu-a-visual-tour_10302234.bc.

The End?

Dave: "Open the pod bay doors, HAL."
HAL: "I'm afraid I can't do that, Dave."

—Lines from 2001: A Space Odyssey, 1968

CHAPTER 9

In the movie, 2001: A Space Odyssey, the main character, Dave is faced with an artificial intelligence, the HAL 9000. In the movie, the AI takes over the spaceship and attempts to kill everyone to preserve some pre-ordained mission. Our last lone human, Dave, simply meticulously crawls into the data bank and takes his trusty screwdriver and removes some sections of memory in order to retake control.

The optimistic message here is that human ingenuity (and a tool) can still save us. Perhaps, he also needed some duct tape to make the toolbox complete. It's hard to write a textbook about scientific development; by the time the book is published, it may well contain newly obsolete ideas.

There is another issue as well, and that is the idea of a "war" on science. We have moved from a period of history where scientific development meant a renaissance of thinking and belief. In our earlier passages about Jules Verne, you could see that Verne believed that a scientist would be the answer to any and everything. People looked at airplanes and electricity as amazing miraculous events. Most people probably expected this trajectory of amazing science to catapult us into an even more amazing future.

So where are our motels on the moon? Where are our flying skateboards? Where is our jetpack? Clearly innovation and new scientific discoveries are not the only issues. Any product or idea must first be vetted not only scientifically but also in the court of public opinion. Why don't we have an outpost on Mars? Perhaps it is because we decided to spend money elsewhere. Why don't we have cloned humans? Perhaps it is because such an idea is anathema to some religious belief structures.

The human condition is not a simply structure. People believe things for a variety of reasons. A cell phone might be the most amazing thing ever when you are a child—and the most essential item currently for you as an adult. But for an elderly person, it can be a miasma for poor sight or loss of hearing. We can argue vociferously against the rights of others because we feel threatened by things we cannot control. Some have begun to distrust the material in front of them since we have access to so much. Polarization is a fact of life. For some reason, giving someone complete and inerrant scientific "proof" is simply not going to change a mind. No longer do we bask in the glory of science. You have friends and you have Facebook friends that you connect with on superficial levels. You may no longer know how to write words because there is no need. Perhaps the future will mean no more simple math. Why learn when the computer can always do it and said computer might be embedded in us for that very purpose?

You don't want a vaccine? It is possible that a miniscule nanovaccine can be embedded in you and it becomes a moot point. Cloning and genetic engineering might mean a future free of disease, but there's probably going to be a few disasters along the way. Improving or changing genes has some long-term ramifications.

Do you need to learn facts? If information is instantly and readily available, what do you need to memorize? Education must develop new methods of evaluation, not just rote learning. If you read something written by a computer, will you know (or care)? Are we still foolishly trying to train our machines to behave the way we want? Or is the machine training us?

Perhaps we will eliminate disease, but the hardy viruses seem to mutate as fast as we can prevent them. Perhaps, that too, will be a thing of the past. But perhaps not. Perhaps human beings will choose to ignore science more and more. Or perhaps we will all join with the machines and find a form of transcendence.

In her short story, *The Ones Who Walk Away From Omelas*, Ursula K. Le Guin tells a story of a beautiful and shining place where everything is perfect and glorious. It is perfection, all is peaceful, no kings, no government, nothing but calm and beauty. In order to keep it that way, though, one child must always live in filthy perpetual darkness. When you are old enough, you learn that truth. Most continue as before. In the Movie *Truman,* Christof, the designer of the place where Truman lives, says, "We accept the reality of the world with which we're presented. It's as simple as that."

Is this modern-day evolution? Only time will tell.

www.ingramcontent.com/pod-product-compliance
Lightning Source LLC
Chambersburg PA
CBHW081540220326
41598CB00036B/6502